THE VARIETIES
of
SCIENTIFIC EXPERIENCE

THE VARIETIES

of

SCIENTIFIC EXPERIENCE

A Personal View of the Search for God

CARL SAGAN

Edited by ANN DRUYAN

Illustrations Editor and Scientific Consultant Steven Soter

THE PENGUIN PRESS

NEW YORK

2006

THE PENGUIN PRESS
Published by the Penguin Group
Penguin Group (USA) Inc., 375 Hudson Street, New York, New York 10014, U.S.A. •
Penguin Group (Canada), 90 Eglinton Avenue East, Suite 700, Toronto, Ontario, Canada M4P
2Y3 (a division of Pearson Penguin Canada Inc.) • Penguin Books Ltd, 80 Strand, London
WC2R 0RL, England • Penguin Ireland, 25 St. Stephen's Green, Dublin 2, Ireland (a division of
Penguin Books Ltd) • Penguin Books Australia Ltd, 250 Camberwell Road, Camberwell,
Victoria 3124, Australia (a division of Pearson Australia Group Pty Ltd) • Penguin Books India
Pvt Ltd, 11 Community Centre, Panchsheel Park, New Delhi - 110 017, India • Penguin Group
(NZ), Cnr Airborne and Rosedale Roads, Albany, Auckland 1310, New Zealand (a division of
Pearson New Zealand Ltd) • Penguin Books (South Africa) (Pty) Ltd, 24 Sturdee Avenue,
Rosebank, Johannesburg 2196, South Africa

Penguin Books Ltd, Registered Offices:
80 Strand, London WC2R 0RL, England

First published in 2006 by The Penguin Press,
a member of Penguin Group (USA) Inc.

Frontispiece figure caption by Ann Druyan,
published in *What Is Enlightenment?* magazine

Illustration credits appear on pages 283–84.

LIBRARY OF CONGRESS CATALOGING IN PUBLICATION DATA
Sagan, Carl, 1934–1996.
The varieties of scientific experience : a personal view
of the search for God / Carl Sagan ; edited by Ann Druyan.
p. cm.
The author's 1985 Gifford lectures.
Includes bibliographical references and index.
Contents: Nature and wonder: a reconnaissance of heaven—The retreat from
Copernicus—The organic universe—Extraterrestrial intelligence—Extraterrestrial
folklore: implications for the evolution of religion—The God hypothesis—The religious experi-
ence—Crimes against creation—The search for who we are—Selected Q&A.
ISBN 1-59420-107-2
1. Natural theology. 2. Religion and science. 3. Sagan, Carl, 1934–1996—Religion.
I. Druyan, Ann, 1949– II. Title.
BL183.S24 2006
215—dc22   2006044827

Printed in the United States of America

1 3 5 7 9 10 8 6 4 2

DESIGNED BY AMANDA DEWEY

Contents

* * *

Carl Sagan was a scientist, but he had some qualities that I associate with the Old Testament. When he came up against a wall—the wall of jargon that mystifies science and withholds its treasures from the rest of us, for example, or the wall around our souls that keeps us from taking the revelations of science to heart—when he came up against one of those topless old walls, he would, like some latter-day Joshua, use all of his many strengths to bring it down.

As a child in Brooklyn, he had recited the Hebrew V'Ahavta prayer from Deuteronomy at temple services: "And you shall love the Lord your God with all your heart, with all your soul, with all your might." He knew it by heart, and it may have been the inspiration for him to first ask, What is love without understanding? And what greater *might* do we possess as human beings than our capacity to question and to learn?

The more Carl learned about nature, about the vastness of the universe and the awesome timescales of cosmic evolution, the more he was uplifted.

Another way in which he was Old Testament: He couldn't

live a compartmentalized life, operating on one set of assumptions in the laboratory and keeping another, conflicting set for the Sabbath. He took the idea of God so seriously that it had to pass the most rigorous standards of scrutiny.

How was it, he wondered, that the eternal and omniscient Creator described in the Bible could confidently assert so many fundamental misconceptions about Creation? Why would the God of the Scriptures be far less knowledgeable about nature than are we, newcomers, who have only just begun to study the universe? He could not bring himself to overlook the Bible's formulation of a flat, six-thousand-year-old earth, and he found especially tragic the notion that we had been created separately from all other living things. The discovery of our relatedness to all life was borne out by countless distinct and compelling lines of evidence. For Carl, Darwin's insight that life evolved over the eons through natural selection was not just better science than Genesis, it also afforded a deeper, more satisfying *spiritual* experience.

He believed that the little we do know about nature suggests that we know even less about God. We had only just managed to get an inkling of the grandeur of the cosmos and its exquisite laws that guide the evolution of trillions if not infinite numbers of worlds. This newly acquired vision made the God who created *the* World seem hopelessly local and dated, bound to transparently human misperceptions and conceits of the past.

This was no glib assertion on his part. He avidly studied the world's religions, both living and defunct, with the same hunger for learning that he brought to scientific subjects. He was enchanted by their poetry and history. When he debated religious leaders, he frequently surprised them with his ability to outquote the sacred texts. Some of these debates led to long-standing friendships and alliances for the protection of life.

However, he never understood why anyone would want to separate science, which is just a way of searching for what is true, from what we hold sacred, which are those truths that inspire love and awe.

His argument was not with God but with those who believed that our understanding of the sacred had been completed. Science's permanently revolutionary conviction that the search for truth never ends seemed to him the only approach with sufficient humility to be worthy of the universe that it revealed. The methodology of science, with its error-correcting mechanism for keeping us honest in spite of our chronic tendencies to project, to misunderstand, to deceive ourselves and others, seemed to him the height of spiritual discipline. If you are searching for sacred knowledge and not just a palliative for your fears, then you will train yourself to be a good skeptic.

The idea that the scientific method should be applied to the deepest of questions is frequently decried as "scientism." This charge is made by those who hold that religious beliefs should be off-limits to scientific scrutiny—that beliefs (convictions without evidence that can be tested) are a sufficient way of knowing. Carl understood this feeling, but he insisted with Bertrand Russell that "what is wanted is not the will to believe, but the desire to find out, which is the exact opposite." And in all things, even when it came to facing his own cruel fate—he succumbed to pneumonia on December 20, 1996, after enduring three bone-marrow transplants—Carl didn't want just to believe: He wanted to know.

Until about five hundred years ago, there had been no such wall separating science and religion. Back then they were one and the same. It was only when a group of religious men who wished "to read God's mind" realized that science would be the most powerful means to do so that a wall was needed. These men—among them Galileo, Kepler, Newton, and, much later, Darwin—began to articulate and internalize the scientific method. Science took off for the stars, and institutional religion, choosing to deny the new revelations, could do little more than build a protective wall around itself.

Science has carried us to the gateway to the universe. And yet our conception of our surroundings remains the disproportionate view of the still-small child. We are spiritually and culturally paralyzed, unable to face the vastness, to embrace our lack of centrality and find our actual place in the fabric of nature. We batter this planet as if we had someplace else to go. That we even do science is a hopeful glimmer of mental health. However, it's not enough merely to accept these insights intellectually while we cling to a spiritual ideology that is not only rootless in nature but also, in many ways, contemptuous of what is natural. Carl believed that our best hope of preserving the exquisite fabric of life on our world would be to take the revelations of science to heart.

And that he did. "Every one of us is, in the cosmic perspective, precious," he wrote in his book *Cosmos*. "If a human disagrees with you, let him live. In a hundred billion galaxies you will not find another." He lobbied NASA for years to instruct *Voyager 2* to look back to Earth and take a picture of it from out by Neptune. Then he asked us to meditate on that image and see our home for what it is—just a tiny "pale blue dot" afloat in the immensity of the universe. He dreamed that we might attain a spiritual understanding of our true circumstances. Like a prophet of old, he wanted to arouse us from our stupor so that we would take action to protect our home.

Carl wanted us to see ourselves not as the failed clay of a disappointed Creator but as *starstuff*, made of atoms forged in the fiery hearts of distant stars. To him we were "starstuff pondering the stars; organized assemblages of 10 billion billion billion atoms considering the evolution of atoms; tracing the long journey by which, here at least, consciousness arose." For him science was, in part, a kind of "informed worship." No single step in the pursuit of enlightenment should ever be considered sacred; only the search was.

This imperative was one of the reasons he was willing to get into so much trouble with his colleagues for tearing down the walls that have excluded most of us from the insights and values of science. Another was his fear that we would be unable to keep even the limited degree of democracy we have achieved. Our society is based on science and high technology, but only a small minority among us has even a superficial understanding of how they work. How can we hope to be responsible citizens of a democratic society, informed decision makers regarding the inevitable challenges posed by these newly acquired powers?

This vision of a critically thoughtful public, awakened to science as a way of thinking, impelled him to speak at many places where scientists were not usually found: kindergartens, naturalization ceremonies, an all-black college in the segregated South of 1962, at demonstrations of nonviolent civil disobedience, on the *Tonight* show. And he did this while maintaining a pioneering, astonishingly productive, fearlessly interdisciplinary scientific career.

He was especially thrilled to be invited to give the Gifford Lectures on Natural Theology of 1985 at the University of Glasgow. He would be following in the footsteps of some of the greatest scientists and philosophers of the last hundred years—including James Frazer, Arthur Eddington, Werner Heisenberg, Niels Bohr, Alfred North Whitehead, Albert Schweitzer, and Hannah Arendt.

Carl saw these lectures as a chance to set down in detail his understanding of the relationship between religion and science and something of his own search to understand the nature of the sacred. In the course of them, he touches on several themes that he had written about elsewhere; however, what follows is the definitive statement of what he took pains to stress were only his personal views on this endlessly fascinating subject.

At the beginning of each Gifford Lecture, a distinguished member of the university community would introduce Carl and marvel at the need for still more additional halls to accommodate the overflow audience. I have been careful not to change the meaning of anything Carl said, but I have taken the liberty of editing out those gracious introductory remarks as well as the hundred or more notations on the audio transcripts that merely say "[Laughter]."

I ask the reader to keep in mind at all times that any deficiencies of this book are my responsibility and not Carl's. Despite the fact that the unedited transcripts reveal a man who spoke extemporaneously in nearly perfect paragraphs, a collection of lectures is not exactly the same thing as a book. This is especially true when the Pulitzer Prize–winning author in question never published anything without combing at least twenty or twenty-five iterations of every manuscript for error or stylistic infelicity.

There was plenty of laughter during these lectures, but also the kind of pin-drop silence that comes when the audience and the speaker are united in the thrall of an idea. The extended dialogues in some of the question-and-answer periods capture a sense of what it was like to explore a question with Carl. I attended every lecture, and more than twenty years later what remains with me was his extraordinary combination of principled, crystal-clear advocacy coupled with respect and tenderness toward those who did not share his views.

The American psychologist and philosopher William James gave the Gifford Lectures in the first years of the twentieth century. He later turned them into an extraordinarily influential book entitled *The Varieties of Religious Experience,* which remains in print till this day. Carl admired James's definition of religion as a "feeling of being at home in the Universe," quoting it at the conclusion of *Pale Blue Dot,* his vision of the human future in space. The title of the book you hold in your hands is a tip of the hat to the illustrious tradition of the Gifford Lectures. My variation on James's title is intended to convey that science opens the way to levels of consciousness that are otherwise inaccessible to us; that, contrary to our cultural bias, the only gratification that science denies to us is deception. I hope it also honors the breadth of searching and the richness of insight that distinguished Carl Sagan's indivisible life and work. The varieties of his scientific experience were exemplified by oneness, humility, community, wonder, love, courage, remembrance, openness, and compassion.

In that same drawer where the transcript of these lectures was rediscovered, there was a sheaf of notes intended for a book we never had the chance to write. Its working title was *Ethos,* and it would have been our attempt to synthesize the spiritual perspectives we derived from the revelations of science. We collected filing cabinets' worth of notes and references on the subject. Among them was a quotation Carl had excerpted from Gottfried Wilhelm Leibniz (1646–1716), the mathematical and philosophical genius, who had invented differential and integral calculus independently of Isaac Newton. Leibniz argued that God should be the wall that stopped all further questioning, as he famously wrote in this passage from *Principles of Nature and Grace:*

"Why does something exist rather than nothing? For 'nothing' is simpler than 'something.' Now this sufficient reason for the existence of the universe . . . which has no need of any other reason . . . must be a necessary being, else we should not have a sufficient reason with which we could stop."

And just beneath the typed quote, three small handwritten words in red pen, a message from Carl to Leibniz and to us: "*So don't stop.*"

· ANN DRUYAN
Ithaca, New York
March 21, 2006

Author's Introduction

In these lectures I would like, following the wording of the Gifford Trust, to tell you something of my views on what at least used to be called natural theology, which, as I understand it, is everything about the world not supplied by revelation. This is a very large subject, and I will necessarily have to pick and choose topics. I want to stress that what I will be saying are my own personal views on this boundary area between science and religion. The amount that has been written on the subject is enormous, certainly more than 10 million pages, or roughly 10^{11} bits of information. That's a very low lower limit. And nevertheless no one can claim to have read even a tiny fraction of that body of literature or even a representative fraction. So it is only in the hope that much that has been written is unnecessary to be read that one can approach the subject at all. I'm aware of many limitations in the depth and breadth of my own understanding of both subjects, and so ask your indulgence. Fortunately, there was a question period after each of the Gifford Lectures, in which the more egregious of my errors could be pointed out, and

I was genuinely delighted by the vigorous give-and-take in those sessions.

Even if definitive statements on these subjects were possible, what follows is not such. My objective is much more modest. I hope only to trace my own thinking and understanding of the subject in the hopes that it will stimulate others to go further, and perhaps through my errors—I hope not to have made many, but it was inevitable that I would—new insights will emerge.

· CARL SAGAN
Glasgow, Scotland
October 14, 1985

THE VARIETIES
of
SCIENTIFIC EXPERIENCE

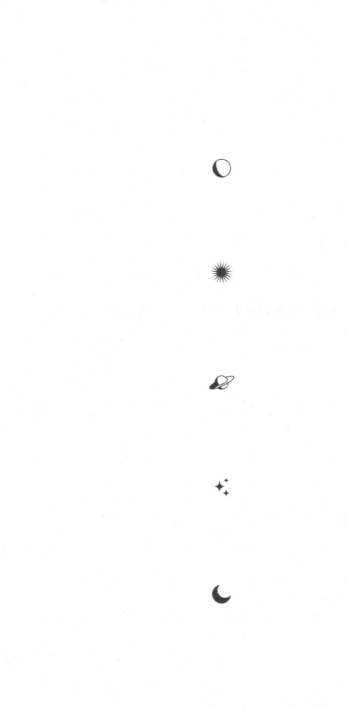

O n e

NATURE AND WONDER:
A RECONNAISSANCE OF HEAVEN

The truly pious must negotiate a difficult course between the
precipice of godlessness and the marsh of superstition.

• Plutarch •

Certainly both extremes are to be avoided, except what are they? What is godlessness? Does not the concern to avoid the "precipice of godlessness" presuppose the very issue that we are to discuss? And what exactly is superstition? Is it just, as some have said, other people's religion? Or is there some standard by which we can detect what constitutes superstition?

For me, I would say that superstition is marked not by its pretension to a body of knowledge but by its method of seeking truth. And I would like to suggest that superstition is very simple: It is merely belief without evidence. The question of what constitutes evidence in this interesting subject, I will try to address. And I will return to this question of the nature of evidence and the need for skeptical thinking in theological inquiry.

The word "religion" comes from the Latin for "binding together," to connect that which has been sundered apart. It's a very interesting concept. And in this sense of seeking the deepest interrelations among things that superficially appear to be sundered, the objectives of religion and science, I believe, are identical or very nearly so. But the question has to do with the reliability of the truths claimed by the two fields and the methods of approach.

By far the best way I know to engage the religious sensibility, the sense of awe, is to look up on a clear night. I believe that it is very difficult to know who we are until we understand where and when we are. I think everyone in every culture has felt a sense of awe and wonder looking at the sky. This is reflected throughout the world in both science and religion. Thomas Carlyle said that wonder is the basis of worship. And Albert Einstein said, "I maintain that the cosmic religious feeling is the strongest and noblest motive for scientific research." So if both Carlyle and Einstein could agree on something, it has a modest possibility of even being right.

Here are two images of the universe. For obvious reasons they concentrate not on the spaces in which there is nothing but on the locales in which there is something. It would be very dull if I simply showed you image after image of darkness. But I stress that the universe is mainly made of nothing, that something is the exception. Nothing is the rule. That darkness is a commonplace; it is light that is the rarity. As between darkness and light, I am unhesitatingly on the side of light (especially in an illustrated book). But we must remember that the universe is an almost complete and impenetrable darkness and the sparse sources of light, the stars, are far beyond our present ability to create or control. This prevalence of darkness, both factually and metaphorically, is worth contemplating before setting out on such an exploration.

fig. 1

fig. 2

fig. 3

This image is intended for orientation. It is an artist's impression of the solar system, in which the sizes of the objects but not their relative distances are to scale. And you can see that there are four large bodies other than the Sun, and the rest is debris. We live on the third piece of debris from the Sun; a tiny world of rock and metal with a thin patina—a veneer—of organic matter on the surface, a tiny fraction of which we happen to constitute.

This picture was made by Thomas Wright of Durham, who published an extraordinary book in 1750, which he quite properly called *An Original Theory or New Hypothesis of the Universe*. Wright was, among other things, an architect and a draftsman. This picture conveys a remarkable sense, for the first time, of looking at the solar system and beyond, to scale. What you can see here is the Sun, and to scale to the size of the Sun is the distance to the orbit of Mercury. Then the planets Venus, Earth, Mars, Jupiter, and Saturn—the other planets were not known in his time—and then, in a wonderful attempt, here is the solar system, the planets we talked about, all in that central dot and a rosette to represent the cometary orbits known in his time. He did not go very far beyond the present orbit of Pluto. And then he imagined, a large distance away, the nearest star then known, Sirius, around which he did not quite have the courage to put another rosette of cometary orbits. But there was the clear sense that our system and the systems of other stars were similar.

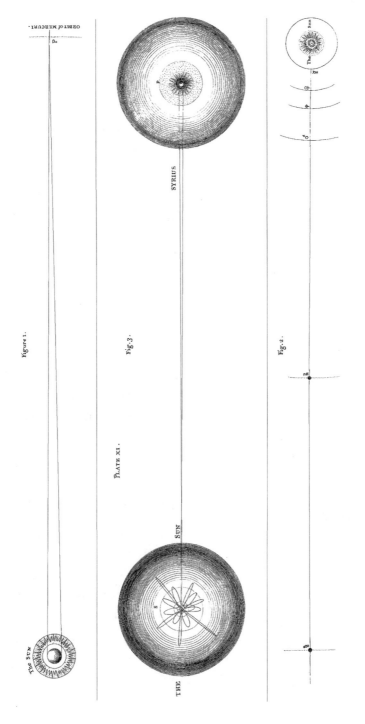

Figure 1.

ORBIT of MERCURY.

P

The Sun

Fig. 3.

PLATE XI.

SYRIUS

SUN

THE

S

Fig. 2.

The Sun

Sun

☿

☿

☊

☋

♂

☿

☿

fig. 4

Here at upper left is the first of four modern illustrations attempting to show just the same thing, in which we see the Earth on its orbit and the other inner planets. Each little dot is intended to represent a fraction of the plethora of small worlds called asteroids. Beyond them is the orbit of Jupiter. And the distance from the Earth to the Sun represented by the scale bar up at the top is called an astronomical unit. This is the first introduction—there will be many of them that I will talk about—of a kind of geocentric or anthropocentric arrogance with which all of the human attempts to look at the cosmos seem to be infected. The idea that an astronomical unit by which we measure the universe has to do with the Earth's distance from the Sun is clearly a human pretension. But since it is deeply embedded in astronomy, I will continue to use the word.

At upper right we see that the previous picture is wrapped in a small square in the middle. Here we have a scale of ten astronomical units. We cannot make out the orbits of the inner planets, including the Earth, on this scale. But we can see the orbits of the giant planets Jupiter, Saturn, Uranus, Neptune, as well as Pluto.

At lower right the previous picture is in a small square, and we now have a scale of a hundred astronomical units. Here's a comet—there are many—with a highly eccentric orbit.

Another increase in scale by a factor of ten and we have the picture at lower left. And here the gray shading is intended to represent the inner boundaries of the Oort Cloud of roughly a trillion comets—cometary nuclei—that surround the Sun and extend to the boundaries of interstellar space.

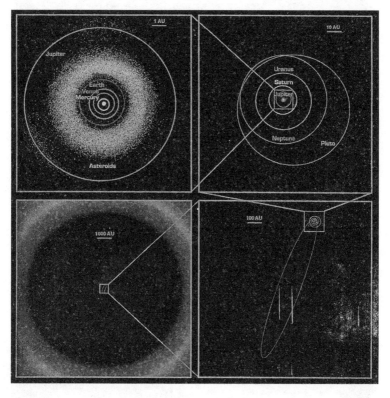

fig. 6

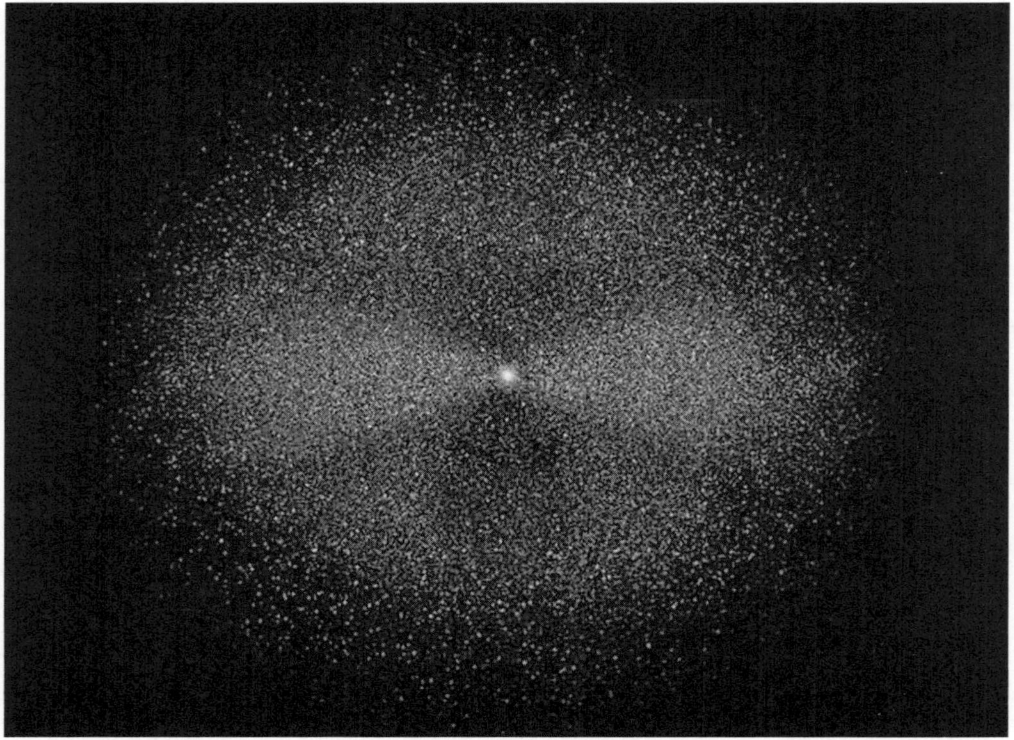

This is an artist's representation of the entire Oort Cloud. Now the dimension is a *hundred thousand* astronomical units, and there is an external boundary to the Oort Cloud. All of the planets, and the comets that we know, are lost in the glare of light from the Sun. And here, for the first time, we have a scale sufficient to see some of the neighboring stars. So the world that we live on is a tiny and insignificant part of a vast collection of worlds, many of which are much smaller, a few of which are much larger. The total number of such worlds are, as I said, something of the order of a trillion, or 10^{12}, a one followed by twelve zeros, of which Earth represents just one, all in the family of the Sun. And our star, of course, is one of a vast multitude.

Here Thomas Wright has made a leap or two, and now we see more than one system with a cometary rosette. He clearly had the sense of the sky being full of systems more or less like our own and was as explicit in words as he is here in a picture in his 1750 book, which, by the way, is also the first explicit statement anywhere that the stars we see in the night sky are part of a concentration of stars that we now call the Milky Way Galaxy, one with a specific shape and a specific center.

There are a vast number of stars within our galaxy. The number is not so large as the number of cometary nuclei around the Sun but is nevertheless hardly modest. It's about 400 billion stars, of which the Sun is one.

fig. 7

fig. 8

This is the Pleiades, a set of young stars that have been born only recently and are still enveloped in their cocoons of interstellar gas and dust.

This is one of the many nebulae, large clouds of interstellar gas and dust. Just to be clear what we are seeing here, there is a sprinkling of foreground stars, behind which is a cloud of glowing interstellar hydrogen—that's the red stuff. The darkness is not the absence of stars; it is simply a place where the dark material prevents you from seeing the stars behind. It is in dense concentrations of this dark interstellar material that new stars and, we now are beginning to see, new planetary systems are in the process of being born.

fig. 9

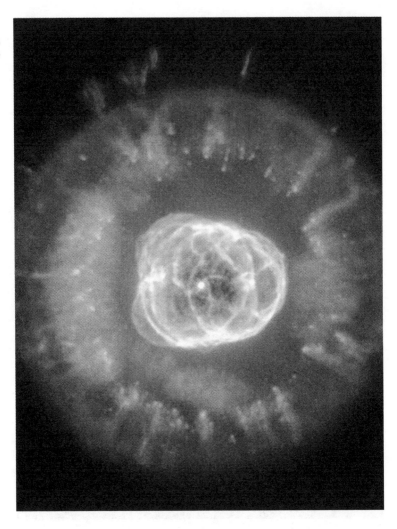

fig. 10

This is a photograph of a dying star. In the course of its evolution, it has expelled its outer layers in a kind of bubble of expanding gas, mainly hydrogen. Stars do this episodically, possibly periodically, and when they do, grave problems are posed for any planets that are around such a star. This is hardly an unusual event for a star a little more massive than the Sun.

Here is a still more explosive and dangerous event. This is the Veil Nebula. It is a supernova remnant, a star that has violently exploded, and any life on any planet that existed around the star that exploded, the supernova, would surely have been destroyed in this explosion. Even ordinary stars like the Sun have a sequence of events late in their history, which mean big trouble for inhabitants of any planets that they might have.

Some 5 or 6 or 7 billion years from now, the Sun will become a red giant star and will engulf the orbits of Mercury and Venus and probably the Earth. The Earth then would be inside the Sun, and some of the problems that face us on this particular day will appear, by comparison, modest. On the other hand, since it is 5,000 or more million years away, it is not our most pressing problem. But it is something to bear in mind. It has theological implications.

fig. 11

fig. 12

There are a huge number of stars. Especially in the center of the galaxy, in the direction of the constellation Sagittarius, the sky is rippling with suns, altogether a couple of hundred thousand million suns, making up the Milky Way Galaxy. As far as we can tell, the average star is in no major way different from the Sun. Or, put another way, the Sun is a reasonably typical star in the Milky Way Galaxy, nothing to call our attention to it. If you had stepped a little bit back and included the Sun in this picture, you would not be able to tell whether it was that one right there or that one right over there, maybe, in the top right-hand corner.

It would be very good to have a photograph of the Milky Way Galaxy taken from an appropriate distance, but we have not yet sent cameras to that distance and so the best we can do for now is to show a photograph of a galaxy like our own, and this is, in fact, the nearest spiral galaxy like our own, M31 in the constellation Andromeda. And again we are looking at stars in the foreground within the Milky Way Galaxy, through which we are seeing M31 and two of its satellite galaxies.

Now, imagine that this is our galaxy. We are looking at a great concentration of stars in the center, so close together that we cannot make out individual ones. We see these spiral lanes of dark gas and dust in which star formation is mainly occurring. If this were the Milky Way Galaxy, where would the Sun be? Would it be in the center of the galaxy, where things are clearly important, or at least well lit? The answer is no. We would be somewhere out in the galactic boondocks, the extreme suburbs, where the action isn't. We are situated in a very unremarkable, unprepossessing location in this great Milky Way Galaxy. But, of course, it is not the only galaxy. There are many galaxies, a very large number of galaxies.

fig. 13

fig. 14

This image is meant to convey just a little sense of how many. We are looking out of the plane of the Milky Way Galaxy in the direction of the Hercules Cluster. What we are seeing here are more galaxies beyond the Milky Way. (In fact, there are more galaxies in the universe than stars within the Milky Way Galaxy.) That is, there are some foreground stars as in the previous pictures, but most of the objects you see here are galaxies—spiral ones seen edge on, elliptical galaxies, and other forms. The number of external galaxies beyond the Milky Way is at least in the thousands of millions and perhaps in the hundreds of thousands of millions, each of which contains a number of stars more or less comparable to that in our own galaxy. So if you multiply out how many stars that means, it is some number—let's see, ten to the . . . It's something like one followed by twenty-three zeros, of which our Sun is but one. It is a useful calibration of our place in the universe. And this vast number of worlds, the enormous scale of the universe, in my view has been taken into account, even superficially, in virtually no religion, and especially no Western religions.

Now, I've not shown you images of our own tiny world, nor did Thomas Wright. He wrote, "To what you have said about my having left out my own habitation in my scheme of the universe, having traveled so far into infinity as but to lose sight of the Earth, I think I may justly answer, as Aristotle did when Alexander, looking over a map of the world, inquired of him for the city of Macedon, 'tis said the philosopher told the prince that the place he sought was much too small to be there taken notice of and was not without sufficient reason omitted. The system of the Sun," Wright goes on, "compared but with a very minute part of the visible creation takes up so small a portion of the known universe that in a very finite view of the immensity of space I judged the seat of the Earth to be of very little consequence."

This perspective provides a kind of calibration of where we are. I don't think it should be too discouraging. It is the reality of the universe we live in.

Many religions have attempted to make statues of their gods very large, and the idea, I suppose, is to make us feel small. But if that's their purpose, they can keep their paltry icons. We need only look up if we wish to feel small. It's after an exercise such as this that many people conclude that the religious sensibility is inevitable. Edward Young, in the eighteenth century, said, "An undevout astronomer is mad," from which I suppose it is essential that we all declare our devotion at risk of being adjudged mad. But devotion to what?

All that we have seen is something of a vast and intricate and lovely universe. There is no particular theological conclusion that comes out of an exercise such as the one we have just gone through. What is more, when we understand something of the astronomical dynamics, the evolution of worlds, we recognize that worlds are born and worlds die, they have lifetimes just as humans do, and therefore that there is a great deal of suffering and death in the cosmos if there is a great deal of life. For example, we've talked about stars in the late stages of their evolution. We've talked about supernova explosions. There are much vaster explosions. There are explosions at the centers of galaxies from what are called quasars. There are other explosions, maybe small quasars. In fact, the Milky Way Galaxy itself has had a set of explosions from its center, some thirty thousand light-years away. And if, as I will speculate later, life and perhaps even intelligence is a cosmic commonplace, then it must follow that there is massive destruction, obliteration of whole planets, that routinely occurs, frequently, throughout the universe.

Well, that is a different view than the traditional Western sense of a deity carefully taking pains to promote the well-being of intelligent creatures. It's a very different sort of conclusion that modern astronomy suggests. There is a passage from Tennyson that comes to mind: "I found Him in the shining of the stars, / I mark'd Him in the flowering of His fields." So far pretty ordinary. "But," Tennyson goes on, "in His ways with men I find Him not. . . . Why is all around us here /As if some lesser god had made the world, / but had not force to shape it as he would . . . ?"

To me personally, the first line, "I found Him in the shining of the stars," is not entirely apparent. It depends on who the Him is. But surely there is a message in the heavens that the finiteness not just of life but of whole worlds, in fact of whole galaxies, is a bit antithetical to the conventional theological views in the West, although not in the East. And this then suggests a broader conclusion. And that is the idea of an immortal Creator. By definition, as Ann Druyan has pointed out, an immortal Creator is a cruel god, because He, never having to face the fear of death, creates innumerable creatures who do. Why should He do that? If He's omniscient, He could be kinder and create immortals, secure from the danger of death. He sets about creating a universe in which at least many parts of it, and perhaps the universe as a whole, dies. And in many myths, the one possibility the gods are most anxious about is that humans will discover some secret of immortality or even, as in the myth of the Tower of Babel, for example, attempt to stride the high heavens. There is a clear imperative in Western religion that humans must remain small and mortal creatures. Why? It's a little bit like the rich imposing poverty on the poor and then asking to be loved because of it. And there are other challenges to the conventional religions from even the most casual look at the sort of cosmos I have presented to you.

Let me read a passage from Thomas Paine, from *The Age of Reason*. Paine was an Englishman who played a major role in both the American and French revolutions. "From whence," Paine asks—"From whence, then, could arise the solitary and strange conceit that the Almighty, who had millions of worlds equally dependent on his protection, should quit the care of all the rest, and come to die in our world because, they say, one man and one woman ate an apple? And, on the other hand, are we to suppose that every world in the boundless creation had an Eve, an apple, a serpent, and a redeemer?"

Paine is saying that we have a theology that is Earth-centered and involves a tiny piece of space, and when we step back, when we attain a broader cosmic perspective, some of it seems very small in scale. And in fact a general problem with much of Western theology in my view is that the God portrayed is too small. It is a god of a tiny world and not a god of a galaxy, much less of a universe.

Now, we can say, "Well, that's just because the right words weren't available back when the first Jewish or Christian or Islamic holy books were written." But clearly that's not the problem; it is certainly possible in the beautiful metaphors in these books to describe something like the galaxy and the universe, and it isn't there. It is a god of one small world, a problem, I believe, that theologians have not adequately addressed.

I don't propose that it is a virtue to revel in our limitations. But it's important to understand how much we do not know. There is an enormous amount we do not know; there is a tiny amount that we do. But what we do understand brings us face-to-face with an awesome cosmos that is simply different from the cosmos of our pious ancestors.

Does trying to understand the universe at all betray a lack of humility? I believe it is true that humility is the only just response in a confrontation with the universe, but not a humility that prevents us from seeking the nature of the universe we are admiring. If we seek that nature, then love can be informed by truth instead of being based on ignorance or self-deception. If a Creator God exists, would He or She or It or whatever the appropriate pronoun is, prefer a kind of sodden blockhead who worships while understanding nothing? Or would He prefer His votaries to admire the real universe in all its intricacy? I would suggest that science is, at least in part, informed worship. My deeply held belief is that if a god of anything like the traditional sort exists, then our curiosity and intelligence are provided by such a god. We would be unappreciative of those gifts if we suppressed our passion to explore the universe and ourselves. On the other hand, if such a traditional god does not exist, then our curiosity and our intelligence are the essential tools for managing our survival in an extremely dangerous time. In either case the enterprise of knowledge is consistent surely with science; it should be with religion, and it is essential for the welfare of the human species.

T w o

·:·

THE RETREAT FROM COPERNICUS: A MODERN LOSS OF NERVE

All of us grow up with the sense that there is some personal relationship between us, ourselves, and the universe. And there is a natural tendency to project our own knowledge, especially self-knowledge, our own feelings, on others. This is a commonplace in psychology and psychiatry. And so it is with our view of the natural world. Anthropologists and historians of religion sometimes call this animism and attribute it to so-called primitive tribes—that is, ones who have not constructed instruments of mass destruction. This is the idea that every tree and brook has a kind of actuating spirit—that, as Thales, the first scientist, said in one of the few surviving fragments of his work, "There are gods in everything." It's a natural idea. But it's not restricted to animists, of whom there are many millions on the planet today. Physicists, for example, do it all the time, except where nature does not oblige. It is the commonest thing in the world in, say, the kinetic theory of gases, to imagine each of these little molecules of air that are busily colliding in front of us as, maybe, billiard balls. Well, that's not exactly projection, since physicists are not strictly speaking of billiard balls, but it is taking something from everyday experience and projecting it into a different realm. It's very common for physicists to refer to molecules or asteroids as "guys." You can more easily imagine what a molecule or an asteroid is like if you imagine them as beings something like us. And this, I believe, reveals the prevalence in this day of these ancient modes of thinking.

Yet you cannot carry this projection too far, because sooner or later you bump your nose. For example, when we get to relativity or quantum mechanics, we discover realms that are alien to our everyday experience, and suddenly the laws of nature turn out to be astonishingly different. The idea that as I walk in this direction my watch goes slightly slower and I am contracted in the direction of motion and my mass has increased slightly does not correspond to everyday experience. Nevertheless, that is an absolutely certain consequence of special relativity, and the reason it does not conform to common sense is that we are not in the habit of traveling close to the speed of light. We may one day be in that habit, and then the Lorentz transformations* will be natural, intuitive. But they aren't yet.

The idea that there is a cosmic speed limit, the speed of light, beyond which no material object can travel, again seems counterintuitive, even though it can be demonstrated, as Einstein did, from an astonishingly simple and basic analysis of what we mean by space, time, simultaneity, and so on.

Or if I were to propose to you that my arm could be in this position or in that position but it would be forbidden by the laws of nature to be in some intermediate position, that would likely

---

*The Lorentz transformations specify how time slows down and length contracts in any frame of reference depending on its relative speed. Einstein's theory of special relativity derived the Lorentz transformation by assuming a constant speed of light for all observers.

strike you as absurd, as contrary to experience. And yet on the subatomic level, there is quantization of energy and position and momentum. The reason it seems counterintuitive is that we are not ordinarily down at the level of the very small, where quantum effects dominate.

So the history of science—especially physics—has in part been the tension between the natural tendency to project our everyday experience on the universe and the universe's noncompliance with this human tendency.

Now, there is another tendency from the psychological or social sphere projected upon the natural world. And that is the idea of privilege. Ever since the invention of civilization, there have been privileged classes in societies. There have been some groups that oppress others and that work to maintain these hierarchies of power. The children of the privileged grow up expecting that, through no particular effort of their own, they will retain a privileged position. At birth all of us imagine that we are the universe, and we don't distinguish the boundaries between ourselves and those around us. This is well established in infants. As we grow up, we discover that there are others who are apparently autonomous and that we're only one among many other people. And then, at least in some social situations, there is the sense that we are central, important. Other social groups, of course, don't have that view. But it is generally those with privilege and status, especially in ancient times, who became the scientists, and there was a natural projection of those attitudes upon the universe.

So, for example, Aristotle provided powerful arguments, none of them instantly dismissible, that the heavens moved and not the Earth, that the Earth is stationary and that the Sun, the Moon, the planets, the stars, rise and set by physically moving once around the Earth every day. With the exception of this kind of motion, the heavens were thought to be changeless. The Earth, while stationary, had all the corruption of the universe localized here.

Up there was matter, which was perfect, unchanging, a special kind of celestial matter that is, by the way, the origin of our word "quintessential." There were four essences down here, the imagined four elements of earth, water, fire, and air, and then there was that fifth element, that fifth essence out of which the heaven stuff was made. And that's why the word "quintessential"—"fifth essence"—comes about. It's interesting to see a kind of linguistic artifact of the previous worldview still present in the *Oxford Unabridged.* But it's amazing what's in the *Oxford Unabridged.*

Now, in the fifteenth century, Nicolaus Copernicus suggested a different view. He proposed that it was the Earth that rotated and that the stars were in effect motionless. He proposed moreover that in order to explain these apparent movements of the planets against the background of more distant stars, the planets and the Earth, in addition to rotating, revolved around the Sun. That is, the Earth was demoted. You know the phrase— another linguistic artifact—*the* world, or *the* Earth. What is the definite article saying? It's saying there is only one. And that also goes straight back to pre-Copernican times, as does the phrase, natural as it is, of the Sun rising and the Sun setting.

Copernicus, by the way, felt his idea to be so dangerous that it was not published until he was on his deathbed, and even then it had an outrageous introduction by a man named Osiander, who was worried that it was too incendiary, too radical. Osiander wrote, in effect, "Copernicus doesn't really believe this. This is just a means of calculating. And don't anybody think he's saying anything contrary to doctrine." This was an important issue.

Aristotle's views had been accepted fully by the medieval church—Thomas Aquinas played a major role in that—and therefore by the time of Copernicus a serious objection to a geocentric universe was a theological offense. And you can see why, because if Copernicus were right, then the Earth would be demoted, no longer *the* Earth, *the* world, but just *a* world, *an* earth, one of many.

And then came the still more unsettling possibility, the idea that the stars were distant suns and that they also had planets going around them and that, after all, you can see thousands of stars with the naked eye. Suddenly the Earth is not only *not* central to this solar system but no longer central to any solar system. Well, there was a period in which we hoped that we were at the center of the Milky Way Galaxy. If we weren't at the center of our solar system, at least our solar system was at the center of the Milky Way Galaxy. And the definitive disproof of that occurred only in the 1920s, to give you an idea of how long it took for Copernican ideas to reach galactic astronomy.

And then there was the hope that, well, at least maybe our galaxy was at the center of all the other galaxies, all those many billions of other galaxies. But modern views have it that there is no such thing as a center of the universe, at least not in ordinary three-dimensional space, and we are certainly not at it.

So those who wished for some central cosmic purpose for us, or at least our world, or at least our solar system, or at least our galaxy, have been disappointed, progressively disappointed. The universe is not responsive to our ambitious expectations. A grinding of heels can be heard screeching across the last five centuries as scientists have revealed the noncentrality of our position and as many others have fought to resist that insight to the bitter end. The Catholic Church threatened Galileo with torture if he persisted in the heresy that it was the Earth that moved and not the Sun and the rest of the celestial bodies. It was serious business.

Now, at the same time, another of the Aristotelian precepts was challenged. That was the idea that except for the moving of crystal spheres into which the planets were embedded, nothing changes up in the heavens. In 1572 there was a supernova explosion in the constellation Cassiopeia. A star that had previously been invisible suddenly became so bright that it could be seen by the naked eye. The Danish astronomer Tycho Brahe noticed it. Well, if nothing changes up there, how is it that suddenly a star appeared—I mean suddenly, in a period of a week or less, from invisibility to something easily seen—and then stayed for some months before fading away? Something was wrong.

Just a few years later, there was an impressive comet, the Comet of 1577, and Tycho Brahe—decades after Copernicus— had the presence of mind to organize an international set of observations of that comet. The idea was to see if it was down here in the Earth's atmosphere, as Aristotle had insisted it must be, or up there among the planets. Part of the reason that Aristotle had insisted that the comets were meteorological phenomena was his belief in an unchanging heaven.

So Brahe thought, if the comet is close to the Earth, then two observers far from each other will see it against different background stars. This is called parallax, which you easily can demonstrate by simply winking your eye, first the left and then the right, with a finger propped up about a foot in front of your nose. The finger seems to move as you blink.

Brahe reasoned that if the comet was very far away, then the two observers who were far apart would see it in almost exactly the same part of the sky. You could determine how far away it was by how much it moved between those two different vantage points, how much the parallax was. And Brahe determined it was surely farther away than the Moon and, therefore, up there, in the planetary realm, and not down here, where the weather is. That was another upsetting discovery for the institutionalized Aristotelian wisdom.

Now, as science has progressed, there have been—one after another—a series of assaults on human vainglory. One of them, for example, is the discovery that the Earth is much older than anyone had expected. Human history goes back only a few thousand years. Many people believed that the world was not much older than human history. And there was no sense of evolution, no sense of vast vistas of time. And then the geological and paleontological evidence began to accumulate, making it very difficult to see how the geological forms and the fossils of now-extinct plants and animals could have come into being, unless the Earth were enormously older than the few thousand years that had been projected. That is a battle still being fought. In the United States, for example, there are people who are called "creationists," the more radical of whom insist that the Earth is less than ten thousand years old. The shorter the age of the Earth, the greater the relative role of humans in the history of the Earth is. If the Earth is, as we certainly know it to be, 4,500 million years old and the human species at most a few million years old, probably less than that, then we have been here for only an instant of geological time, for less than one one-thousandth of the history of the Earth, and therefore in time, as in space, we have been demoted from the central to an incidental aspect.

And then evolution itself was still a further disquieting discovery, because at least it had been hoped that humans were separate from the rest of the natural world, that we had been specifically put here in a way different from petunias, let's say. And yet Darwin's historic work showed that we were very likely related in an evolutionary sense with all the other beasts and vegetables on the planet. And there remain many people who are enormously offended by this idea.

This sense of offense has—I'm only speculating—deep psychological roots. Part of it is, I believe, an unwillingness to come to grips with the more instinctive aspects of human nature. But I believe it is essential to understand this if we wish to survive. I think ignoring that, imagining all humans are rational actors in the present phase, is enormously dangerous in an age of nuclear weapons. I think the discomfort that some people feel in going to the monkey cages at the zoo is a warning sign.

Then, in the early part of this century, there was still another such assault, which came with special relativity. Because one of the central points of special relativity is that there are no privileged frames of reference, that we are not in an important position or state of motion. There is nothing privileged about the velocity that we have or the acceleration that we have; the universe can be understood precisely if it is true that we do not have a special frame of reference.

Now, it's certainly true that there is something special about our position in time. The universe has changed. A microsecond after the Big Bang, it was quite different from how it is right now. So no one maintains these days that there is not something special about our epoch in the sense that the universe itself evolves. But in terms of position, velocity, and acceleration, there is nothing privileged about where we are. This insight was obtained by a young man who was opposed to privilege in the social sphere. If you look at Einstein's autobiographical writings, I think it is quite clear that his opposition to privilege in the social world was connected with his opposition to privilege in fundamental physics.

Well, if we don't have a distinctive position or velocity or acceleration, or a separate origin from the other plants and animals, then at least, maybe, we are the smartest beings in the entire universe. And that is our uniqueness. So today the battle, the Copernican battle, is, in somewhat covert form, being waged on the issue of extraterrestrial intelligence. Now, this doesn't guarantee that there is extraterrestrial intelligence. It may be that the Copernican insights—the principle of mediocrity, if you wish to call it that—worked for all these other things, but on extraterrestrial life it doesn't, and we are unique. I will come back to that later, but I believe that the ongoing Copernican revolution is relevant to this debate as well.

There is today another battlefield on which the Copernican insights are being attacked. It is connected with one of the classic arguments for the existence of God, that is, the Western kind of God, namely, the argument from design.

⁙

The idea of the argument from design goes like this: Suppose you know nothing about watches and you come upon an elegantly tooled pocketwatch. And you open it and everything is going *tick-tick-tick-tick*, and there are all those gears and levers and burnished brass, and such things are not made in nature. Therefore the existence of such a complex mechanism, the existence of the watch, implies a watchmaker. Now we go and look at an organism. Let's take a very modest organism, a bacterium. Well, you look in there and you find a much more complex mechanism than a pocketwatch. A bacterium has many more moving parts, much more information than what you would have to write down in order to describe how to make a pocketwatch. And yet the world is full of bacteria. They're everywhere, enormous quantities of them. And is it possible that this being, far more complex than a watch, arose spontaneously out of who knows what sort of collisions of atoms? Isn't it more likely that this "watch" also implies a watchmaker? That is one example of the argument from design, and you can imagine that every part of nature might be vulnerable to such an interpretation. Everything, that is, except utter chaos.

Well, Darwin showed, through natural selection, that there was a way other than the existence of a Watchmaker, a way in which it was possible for enormous order to emerge from a more disordered natural world without the interposition of any capital-*W* Watchmaker. That was natural selection.

The ideas behind natural selection were that there was such a thing as a hereditary material, that there were spontaneous changes in the hereditary material, that those changes were expressed in the external form and function of the organism, that organisms made many more copies of themselves than the environment could support, and therefore that some selection among various natural experiments was made by the environment for reproductive success, that some organisms, by pure accident, were better suited to leaving offspring than others.

Now, an essential aspect of this idea is that you need to have enough time. If the universe is only a few thousand years old, then Darwinian evolution is nonsense. There isn't time. On the other hand, if the Earth is a few thousand million years old, then there is enormous time, and we can at least contemplate that this is the source, as certainly all of modern biology suggests, of the complexity and beauty of the biological world.

This sort of argument from design we can find in other aspects of nature. And I'd like to discuss two of them. One is Isaac Newton's understanding of the order within the solar system, and the other is a most interesting although, I believe, flawed approach to the laws of nature, put forth in recent times, called the "anthropic principle."

One of Newton's many extraordinary accomplishments was to show that, granting a few simple and highly nonarbitrary laws of nature, he could deduce to high precision the motion of the planets in the solar system. The Newtonian method has remained valid from that time to this. It is precisely Newtonian physics that is used routinely in my line of work, sending spacecraft to the planets, something that you might be tempted to say was far beyond Newton's expectations. But he in fact did envision at least the launching of objects into Earth orbit.

What Newton found is that there is a distinctive plane to the solar system. Copernicus had essentially proposed this, but Newton showed in detail how it worked. The orbits of the planets circle the Sun, all of them very close to the ecliptic plane, also called the zodiacal plane (because the constellations of the zodiac are arrayed around this plane). And that's why the planets and the Sun and the Moon apparently move through the zodiac. "Why is everything so regular?" Newton asked. "Why are all the planets in the same plane? Why do they all go around the Sun in the same direction?" It's not that Mercury goes around one way while Venus goes around in another way. All of the planets go around in the same sense. And, as far as he knew then, they all rotated in the same sense. The planets had something astonishingly regular about them. On the other hand, the comets that were known in his day were helter-skelter. Their orbits were at every possible angle to the ecliptic plane. Some went around in the direct sense; some went around in the retrograde sense. And they were tilted in all sorts of directions.

Newton believed that the distribution of cometary orbits was the state of nature and that is how the planets would have moved had there not been an intervening hand. He believed that God established the initial conditions for the planets that made them all go around the Sun in the same direction, in the same plane, and rotating in a compatible sense.

Now, in fact, this is not a strong conclusion. And Newton, who was extraordinarily perceptive in so many areas, was clearly not here.

The outline of a general solution of this problem was provided, independently as far as we can tell, by both Immanuel Kant and by Pierre-Simon, the marquis de Laplace.

Newton, Laplace, and Kant all lived after the invention of the telescope and therefore after the discovery that Saturn has an exquisite ring system that goes around it, a portion of which you see here in this far-encounter photograph. It is a flat plane of clearly fine particles. The first clear demonstration that it's made of many particles, that it isn't a solid sheet, was made by a Scottish physicist, James Clerk Maxwell.

Here's a closer view of the rings of Saturn. And you can see an enormous sequence of such rings and a gap—the so-called Cassini Division in the rings.

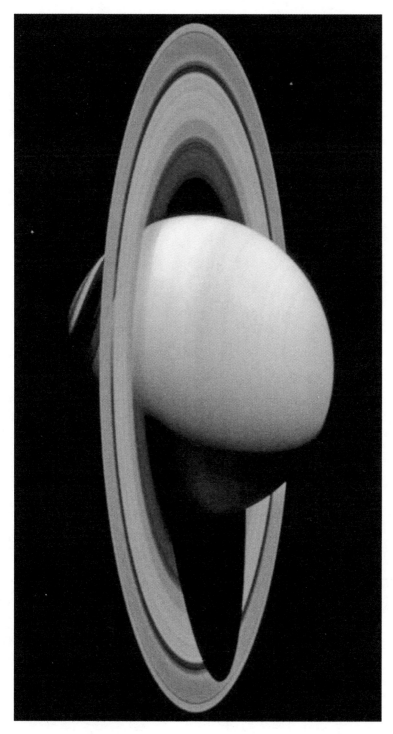

fig. 15

fig. 16

If you take a close-up look at this portion, you can see a succession of rings. We now know that there are many hundreds of these rings, all in a flat plane, and we now know, as both Kant and Laplace guessed, that they're made of tumbling boulders and dust particles. The rings of Saturn, by the way, are thinner compared to their lateral extent than is a piece of paper.

Kant also knew about objects that were then called nebulae. It was not clear whether they were within our Milky Way or beyond—we now know, of course, most of them are beyond. Some of the nebulae were again flattened systems made, we now know, of stars.

So Kant and Laplace, both of them explicitly mentioning the rings of Saturn, and Kant explicitly mentioning the elliptical nebulae, proposed that the solar system came from such a flattened disk and that somehow the planets condensed out of the disk. But if that's the case, the disk, after all, has some rotation. Everything that condenses out of it will be going around in the same direction. And if you think about it for a moment, you will see that as the particles come together and make larger objects, they will have a common sense of rotation as well.

What Kant and Laplace proposed is what we now call a solar nebula, or accretion disk, whose flattened form was the ancestor of the planets, and that it is perfectly easy to understand how it is that the planets are in the same plane with the same direction of revolution and the same sense of rotation.

What is more, we now know that the random orientation of the comets is not primordial and that very likely the comets began in the solar nebula, all going around the Sun in the same sense, were ejected by gravitational interactions with the major planets, and then, by the gravitational perturbations of passing stars, had their orbits randomized.

So Newton was wrong in both senses: (a) in the sense of believing that the chaotic distribution of cometary orbits is what you would expect in a primordial system and (b) in assuming that there was no natural way in which the regularities of planetary motion could be understood without divine intervention, from which he deduced the existence of a Creator.

Well, if Newton could be fooled, this is something worth paying attention to. It suggests that we, of doubtless inferior intellectual accomplishment, might be vulnerable to the same sort of error.

I would just like to lock in what I've been saying about the solar nebula with three more images.

Here is an attempt to illustrate what I've just been saying. An originally irregular interstellar cloud is rotating. It gravitationally contracts; that is, the self-gravity pulls it in. Because of the conservation of angular momentum, it flattens into a disk. A way to think of it is that centrifugal force does not oppose the contraction along the axis of rotation but does in the plane of rotation. So you can see that the net result will be a disk. Through processes that need not detain us here (although remarkable progress has been made in our understanding during the last twenty years), there are gravitational instabilities that produce a large number of objects, which then fall together by collision and produce a smaller number of objects. It's clear that

fig. 17

if there were a huge number of objects with crossing orbits, they would eventually collide, and you would wind up with fewer and fewer objects. So the idea here is that there is a kind of collisional natural selection—the evolutionary idea as applied to astronomy—in which you must eventually wind up with a small number of objects in orbits that do not cross each other. And that is certainly the present configuration of the planetary system shown up here.

This is just another artist's conception of an early stage in the origin of our solar system, showing some of the multitude of small objects a few kilometers across, from which the planets were formed. And that this is not solely a theoretical construct has been made clear in recent years by the discovery of a number of flattened disks around nearby stars.

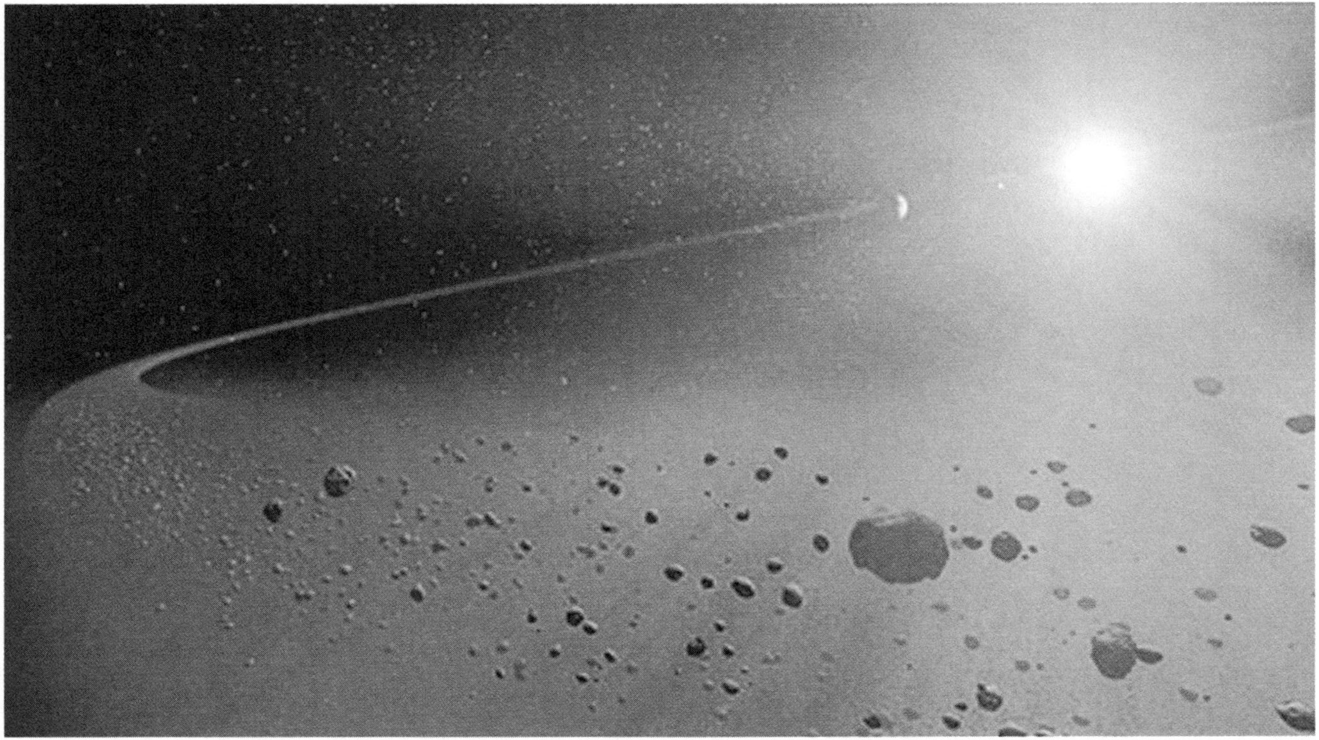

fig. 18

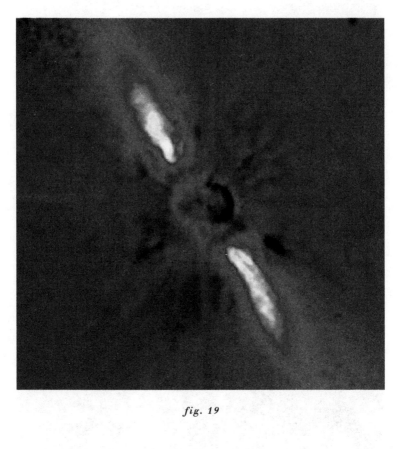

fig. 19

This one is around the star Beta Pictoris. It's in a Southern Hemisphere constellation. But Vega, one of the brightest stars in the Northern sky, also has such a flattened disk of dust and maybe a little gas around it. And many people think that it is in the final stages of sweeping up a solar nebula, that planets have already formed there, and that if you come back in only a few tens of millions of years you will find the disk entirely dissipated and a fully formed planetary system.

So I would like now to come to what is called the anthropic principle. If you study history, it's almost irresistible to ask the question, what if something had gone in a different direction? What if George III had been a nice guy? There are many questions; that's not the deepest, but you understand what I'm saying. There are many such apparently random events that could just as easily have gone another way, and the history of the world would be significantly different. Maybe—I don't know that this is the case—but maybe Napoleon's mother sneezed and Napoleon's father said, "Gesundheit," and that's how they met. And so a single particle of dust was responsible for that deviation in human history. And you can think of still more significant possibilities. It's a natural thing to think about.

Now, here we are. We're alive; we have some modest degree of intelligence; there is a universe around us that clearly permits the evolution of life and intelligence. That's an unremarkable and, I think, as secure a remark as can be made in this subject: that the universe is consistent with the evolution of life, at least here. But what is interesting is that in a number of respects the universe is very fine-tuned, so that if things were a little different, if the laws of nature were a little different, if the constants that determine the action of these laws of nature were a little different, then the universe might be so different as to be incompatible with life.

For example, we know that the galaxies are all running away from each other (the so-called expanding universe). We can measure the rate of expansion (it is not strictly constant with time). We can even extrapolate back and ask how long ago were all the galaxies so close that they were in effect touching. And that will surely be, if not the origin of the universe, at least an anomalous or singular circumstance from which we can begin dating. And that number varies according to a number of estimates, but it's roughly 14,000 million years.

Now, the period of time that was required for the evolution of intelligent life in the universe—if we are unique and we define ourselves immodestly as the carriers of intelligent life (a case could be made, you know, for other primates and dolphins, whales, and so on)—but for any of those cases it took something like 14,000 million years for intelligence to arrive. Well, how come? Why are those two numbers the same? Put another way: If we were at a much earlier stage or a much later stage in the expansion of the universe, would things be very different? If we were at a much earlier stage, then there would not be, according to this view, enough time for the random aspects of the evolutionary process to proceed, and so intelligent life would not be here, and so there would be nobody to make this argument or debate about it. Therefore the very fact that we can talk about this demonstrates, it is argued, that the universe must be a certain number of years old. So if only we had been wise enough to have thought of this argument before Edwin Hubble, we could have made this spectacular discovery about the expansion of the universe just by contemplating our navels.

There is to my mind a very curious ex post facto aspect of this argument. Let's take another example. Newtonian gravitation is an inverse square law. Take two self-gravitating objects, move them twice as far apart, the gravitational attraction is one-quarter; move them ten times farther apart, the gravitational attraction is one-hundredth, and so on. It turns out that virtually any deviation from an exact inverse square law produces planetary orbits that are, in one way or another, unstable. An inverse cube law, for example, and higher powers of the negative exponent mean that the planets would rapidly spiral into the Sun and be destroyed.

Imagine a device with a dial for changing the law of gravity (I wish there were such a device, but there isn't). We could dial in any exponent, including the number 2 for the universe we live in. And when we do this, we find that a large subset of possible exponents leads to a universe in which stable planetary orbits are impossible. And even a tiny deviation from 2—2.0001, for example—might, over the period of time of the history of the universe, be enough to make our existence today impossible.

So, one may ask, how is it that it's exactly an inverse square law? How did it come about? Here is a law that applies to the entire cosmos that we can see. Distant binary galaxies going around each other follow exactly an inverse square law. Why not some other sort of law? Is it just an accident, or is there an inverse square law so that we could be here?

In the same Newtonian equation, there is the gravitational coupling constant called "big G." It turns out that if big G were ten times larger (its value in the centimeter-gram-second system is about 6.67×10^{-8}), if it were 10 times larger (6.67×10^{-7}), then it turns out the only kind of stars we would have in the sky would be blue giant stars, which expend their nuclear fuel so rapidly that they would not persist long enough for life to evolve on any of their planets (that is, if the timescales for the evolution of life on our planet are typical).

Or if the Newtonian gravitational constant were ten times less, then we would have only red dwarf stars. What's wrong with a universe made with red dwarf stars? Well, it is argued, they're around for a long time because they burn their nuclear fuel slowly, but they are such feeble sources of light that to be warmed to the temperatures of liquid water, let's say,* then the planets would have to be very close to the star in order to be at this temperature. But if you put the planets very close to the star, there is a tidal pull that the star exerts on the planet so that the planet always keeps the same face to the star, and therefore, it is said, the near side will be too hot and the far side will be too cold and it's inconsistent with life. So isn't it remarkable that big *G* has the value it does? I'll come back to this.

Or consider the stability of atoms. An electron with something like one eighteen-hundredth the mass of a proton has precisely the same electrical charge. Precisely. If it were even a little different, the atoms would not be stable. How come the electrical charges are exactly the same? Is it so that 14 billion years later we, who are made of atoms, could be around?

Or if the strong nuclear force coupling constant were only a little weaker than it is, you can show that only hydrogen would be stable in the universe and all the other atoms, which surely

---

*There is something anthropocentric without a doubt in talking about liquid water, but let's grant them that. It's curious in these arguments to find organisms who are made largely of liquid water saying that liquid water is central to the universe. But put that aside.

are required for life, we would say, would never have been made.

Or if certain specific nuclear resonances in the nuclear physics of carbon and oxygen were a little different, then you could not build up in the interiors of red giant stars the heavier elements and again you would have only hydrogen and helium in the universe and life would be impossible. How is it that everything works out so well to permit life when it's possible to imagine quite different universes?

(What I'm about to say now is not an answer to the question I've just posed.) It is not difficult to see teleology hiding in this sequence of arguments. And, in fact, the very phrase "anthropic principle" is a giveaway as to at least the emotional if not the logical underpinnings of the argument. It says something central about us; we're the *anthropos*. And that's the reason I am saying that this is another ground, somewhat covert, on which the Copernican conflict is being worked out in our time. J. D. Barrow, one of the authors and promoters of the anthropic principle, is quite straightforward about it. He says that the universe is "designed with the goal of generating and sustaining observers"—namely, us.

Now, what can we say about this? Let me make, in conclusion, a few critical remarks. First of all, in at least parts of this argument there is a failure of the imagination. Let's take that red dwarf argument, in which if the gravitational constant were an order of magnitude less, then we would only have those red giants. Is it true that you could not have life in that situation for the reasons I mentioned? It turns out it isn't, for two different reasons. Let's look again at that tidal locking argument. Yes, for a close-in planet and the star, it seems possible that the net result would be the same kind of situation as for the Moon and the Earth, namely, that the secondary body makes one rotation per revolution, therefore always keeping the same face to the primary. That's why we always just see one Man in the Moon and not some Woman in the Moon on the back that we see as well. But if you look at Mercury and the Sun, you find a close-in planet not in a one-to-one resonance, but it's a three-to-two resonance. There are many more than just this one kind of resonance that are possible. What is more, if we're talking about planets that have life, we're talking about planets with atmospheres. A planet with an atmosphere carries the heat from the illuminated to the unilluminated hemisphere and redistributes the temperature. So it's not just the hot side and the cold side. It is much more moderate than that.

And then let's take a look at the more distant planets that you might imagine were too cold to support life. This neglects what is called the greenhouse effect, the keeping in of infrared emission by the atmospheres of the planet. Let's take Neptune, at thirty astronomical units from the Sun, so you would figure that it has almost a thousand times less sunlight. And yet there is a place we can see with radio waves in the atmosphere of Neptune that is as warm as it is in the cozy room I'm in. So what has happened here is that an argument has been put forward, but in insufficient detail. It has not been looked at closely enough. And I bet that will turn out to be the case in some of the other examples I present.

The second possibility is that there is some new principle hitherto undiscovered, which connects various apparently unconnected aspects of the universe in the same way that natural selection provided a wholly unexpected solution to a problem that seemed to have no conceivable solution whatever.

And thirdly, there is the so-called many worlds or, better,

many universes idea. And this is what I had in mind when I was talking about history at the beginning. Namely, that if at every microinstant of time the universe splits into alternate universes in which things go differently, and that if there is at the same moment an enormously, tremendously large, perhaps infinitely large array of other universes with other laws of nature and other constants, then our existence is not really that remarkable. There are all those other universes in which there isn't any life. We just, by accident, happen to be in the one that has life. It's a little bit like a winning hand at bridge. The chance of, let's say, being dealt twelve spades is an absurdly low probability. But it is as likely as getting any other hand, and therefore, eventually, if you play long enough, some universe has to have our laws of nature.

Well, I believe that we are seeing a still largely unexplored area of physics being projected upon by the same sorts of human hopes and fears that have characterized the entire history of the Copernican debate.

I wanted to say just two final things. One is, if the very strong version of the anthropic principle is true, that is, that God—we might as well call a spade a spade—created the universe so that humans would eventually come about, then we have to ask the question, what happens if humans destroy themselves? That would make the whole exercise sort of pointless. So if only we could believe the strong version, we would have to conclude either (a) that an omnipotent and omniscient God did not create the universe, that is, that He was an inexpert cosmic engineer, or (b) that human beings will not self-destruct. Either alternative, it seems to me, is a matter of some interest, would be worth knowing. But there is a dangerous fatalism lurking here in the second branch of that fork in this road.

Well, I would like to conclude, then, by just a few lines of poetry, this one from Rupert Brooke, called "Heaven."

FISH (fly-replete, in depth of June,
Dawdling away their wat'ry noon)
Ponder deep wisdom, dark or clear,
Each secret fishy hope or fear.

Fish say, they have their Stream and Pond;
But is there anything Beyond?
This life cannot be All, they swear,
For how unpleasant, if it were!

One may not doubt that, somehow, Good
Shall come of Water and of Mud;
And, sure, the reverent eye must see
A Purpose in Liquidity.

We darkly know, by Faith we cry,
The future is not Wholly Dry.
Mud unto mud!—Death eddies near—
Not here the appointed End, not here!

But somewhere, beyond Space and Time,
Is wetter water, slimier slime!
And there (they trust) there swimmeth One,
Who swam ere rivers were begun,

Immense, of fishy form and mind,
Squamous, omnipotent, and kind;
And under that Almighty Fin,
The littlest fish may enter in.

Oh! never fly conceals a hook,

Fish say, in the Eternal Brook,

But more than mundane weeds are there,

And mud, celestially fair;

Fat caterpillars drift around,

And Paradisal grubs are found;

Unfading moths, immortal flies,

And the worm that never dies.

And in that Heaven of all their wish,

There shall be no more land, say fish.

Three

∴

THE ORGANIC UNIVERSE

Once upon a time, the best minds of the human species believed that the planets were attached to crystal spheres,
which explained their motion both daily and over longer periods of time. We now know this is not true in several ways, one of
which is that the Copernican theory explains the observed motion to higher precision and with a more modest investment of
assumptions. But we also know this is not true, because we have
sent spacecraft to the outer solar system with acoustic micro-
meteorite detectors—and there was no sound of tinkling crystal
as the spacecraft passed the orbits of Mars or Jupiter or Saturn.
We have direct evidence that there are no crystal spheres. Now,
Copernicus did not have such evidence, of course, but nevertheless his more indirect approach has been thoroughly validated.
Now, when they were believed to exist, how was it that these
spheres moved? Did they move on their own? They did not.
Both in classic and in medieval times, it was prominently speculated that gods or angels propelled them, gave them a twirl
every now and then.

The Newtonian gravitational superstructure replaced angels with GMm/r^2, which is a little more abstract. And in the course of that transformation, the gods and angels were relegated to more remote times and more distant causality skeins. The history of science in the last five centuries has done that repeatedly, a lot of walking away from divine microintervention in earthly affairs. It used to be that the flowering of every plant was due to direct intervention by the Deity. Now we understand something about plant hormones and phototropism, and virtually no one imagines that God directly commands the individual flowers to bloom.

So as science advances, there seems to be less and less for God to do. It's a big universe, of course, so He, She, or It could be profitably employed in many places. But what has clearly been happening is that evolving before our eyes has been a God of the Gaps; that is, whatever it is we cannot explain lately is attributed to God. And then after a while, we explain it, and so that's no longer God's realm. The theologians give that one up, and it walks over onto the science side of the duty roster.

We've seen this happen repeatedly. And so what has happened is that God is moving—if there is a real God of the Western sort, I am, of course, speaking only metaphorically— God has been evolving toward what the French call *un roi fainéant*—a do-nothing king—who gets the universe going, establishes the laws of nature, and then retires or goes somewhere else. This is not far at all from the Aristotelian view of the unmoved prime mover, except that Aristotle had several dozen unmoved prime movers, and he felt that this was an argument for polytheism, something that is often overlooked today.

Well, I want to describe one of the most major gaps that is in the course of being filled in. (We cannot surely say it is fully filled in yet.) And that has to do with the origin of life.

There was, and in some places still is, a very intense controversy about the evolution of life, about the scandalous suggestion that humans are closely related to the other animals and especially to nonhuman primates, that we had an ancestor who would be, if we met it on the street, indistinguishable from a monkey or an ape. A great deal of the attention has been devoted to the evolutionary process, where, as I tried to indicate earlier, the key impediment to its being intuitively obvious is time. The period of time available for the origin and evolution of life is so much vaster than an individual human lifetime that processes that proceed at paces too small to see during an individual lifetime might nevertheless be dominant over 4,000 million years.

One way to think about this, by the way, is the following: Suppose your father or mother—let's say father for the sake of definiteness—walked into this room at the ordinary human pace of walking. And suppose just behind him was *his* father. And just behind him was *his* father. How long would we have to wait before the ancestor who enters the now-open door is a creature who normally walked on all fours? The answer is a week. The parade of ancestors moving at the ordinary pace of walking would take only a week before you got to a quadruped. And our quadruped ancestors are, after all, only a few tens of millions of years ago, and that's 1 percent of geological time. So there are many different ways of calibrating this immense vista of time that was necessary to evolve the complexity and beauty of the natural world, and this is one.

Now, the evidence for evolution is ubiquitous, and I will not spend a great deal of time on it here. But just to remind everyone. The centerpiece is, of course, the fossil record. Here we find a correlation of geological strata otherwise identifiable and datable by radioactive dating and other methods—with fossils, the remains, the hard parts—of organisms largely now extinct.

If you looked at an undisturbed sedimentary column, the remains of human beings would be found only in the very topmost layers. The farther down you dig, the farther back in time you are going. And no one has ever found any remnant of a human being down in the Jurassic or the Cambrian or any of the geological time periods other than the most recent—the last few million years. And likewise there are many organisms that were absolutely dominant and abundant worldwide for enormous periods of time that became extinct and were never seen again in the higher sedimentary columns. Trilobites are an example. They hunted in herds on the ocean bottoms. They were enormously abundant, and there have not been any of them on the Earth since the Permian. In fact, by far most of the species of life that have ever existed are now extinct. Extinction is the rule. Survival is the exception.

When you look at the fossil record, it is clear that some organisms have powerful anatomical similarities with others. Others are more distinct. There is a kind of taxonomic evolutionary tree that has been painstakingly developed over a century or more. But in recent times it is possible to look for chemical fossils—to examine the biochemistry of organisms that are alive today—and we are even just beginning to know something about the biochemistry of organisms that are extinct, because some of their organic matter can nevertheless be recovered. And here there is a remarkable correlation between what the anatomists say and the molecular biologists say. So the bone structure of chimpanzees and humans is startlingly similar. And then you look at their hemoglobin molecules, and they are startlingly similar. There's only one amino acid difference out of hundreds between the hemoglobins of chimps and humans.

In fact when you look more generally at life on Earth, you find that it is all the same kind of life. There are not many different kinds; there's only one kind. It uses about fifty fundamental biological building blocks, organic molecules. (By the way, when I use the word "organic," there is no necessary implication of biological origin. All I mean when I say organic is a molecule based on carbon that's more complicated than CO and CO_2.)

Now, it turns out that with trivial exceptions all organisms on Earth use a particular kind of molecule called a protein as a catalyst, an enzyme, to control the rate and direction of the chemistry of life. All organisms on Earth use a kind of molecule called a nucleic acid to encode the hereditary information and to reproduce it in the next generation. All organisms on Earth use the identical code book for translating nucleic acid language into protein language. And while there are clearly some differences between, say, me and a slime mold, fundamentally we are tremendously closely related. The lesson is, don't judge a book by its cover. At the molecular level, we are all virtually identical.

This then raises interesting questions about whether we have any idea of the possible range of life, of what could be elsewhere. We are trapped in a single example and have not the imagination to guess even one other way in which life might exist when there might be thousands or millions. Certainly no one deduced from fundamental theoretical chemistry the existence and function of nucleic acids when they were all around us and, in fact, when we ourselves were made of them.

Now, how did it come about that these few particular molecules, out of the enormous range of possible organic molecules, determine all life on Earth? There are two main possibilities and a range of intermediate cases. One possibility is that these molecules were somehow made preferentially in great abundance in the early history of the Earth, and so life just used what was lying around.

The other possibility is that these molecules have some special properties that are not only germane but essential for life, and so they were gradually developed by living systems or preferentially removed from a dilute to a concentrated solution by them. And, as I said, there is a range of intermediate possibilities.

It would be wrong to say that the origin of proteins and nucleic acids is identical with the origin of life. And yet nucleic acids are known in the laboratory to replicate themselves and even to replicate changes in themselves from plausible building blocks in the medium. It is true that an enzyme is needed for this reaction in the laboratory, but this enzyme determines the rate and not the direction of the chemical reaction, so it merely shows us what would happen were we willing to wait long enough. And there was surely plenty of time for the origin of life, which I will come back to as well.

It is certainly conceivable that what we have today is quite different from what was present at the time of the origin of life. We have today a very sophisticated kind of life, evolved by natural selection, that was based upon something much simpler, much earlier. It has been proposed that "much simpler" might in fact be mainly inorganic or it may have been organic; there is no way to be sure. But one thing is undoubtedly of interest for the origin of life—some would say essential—and that is to understand where the molecular building blocks that are present in all living things today came from.

So we now come to the issue of organic molecules. They are found on the Earth, of course, but since the Earth is littered with life, we do not have a clean experiment. We don't know, or at least it's not immediately obvious, which organic molecules we see on the Earth are here because of life and which would be here even if there had not been life. And virtually all the organic molecules that we see in our everyday lives are of biological origin. If you want to know something about organic chemistry on the Earth prior to the origin of life, it is a good idea to look elsewhere.

The idea of extraterrestrial organic matter is important not just for this reason but also because it tells us something relevant at least about the likelihood of extraterrestrial life. If it turns out that there is no sign of organic molecules elsewhere, or they're extremely rare, that might lead you to conclude that life elsewhere was extremely rare. If you found the universe burgeoning and overflowing with organic matter, then at least that prerequisite for extraterrestrial life would be satisfied. So it's an important issue. It's an issue where remarkable progress has been made since the early 1950s, and it speaks to us, I believe, if not centrally at least tangentially, about our origins.

The astronomer Sir William Huggins frightened the world in 1910. He was minding his own business, doing astronomy, but as a result of his astronomy (the work I'm talking about was done in the last third of the nineteenth century) there were national panics in Japan, in Russia, in much of the southern and midwestern United States. A hundred thousand people in their pajamas emerged onto the roofs of Constantinople. The pope issued a statement condemning the hoarding of cylinders of oxygen in Rome. And there were people all over the world who committed suicide. All because of Sir William Huggins's work. Very few scientists can make similar claims. At least until the invention of nuclear weapons. What exactly did he do? Well, Huggins was one of the first astronomical spectroscopists.

fig. 20

This is the coma of a comet—the cloud of gas and dust that surrounds the icy comet nucleus when it enters the inner solar system. Huggins used a spectroscope to spread out the light from a comet into its constituent frequencies. Some frequencies of light are preferentially present, from which it is possible to deduce something of the chemistry of the material in the comet. This is an application of stellar spectroscopy that had been going very successfully in the decade or two before Huggins turned his attention to the comets. (Huggins also made major contributions to understanding the chemistry of the stars.)

This image of four spectra is taken from one of Huggins's publications. These are wavelengths of light in the visible part of the spectrum to which the eye is sensitive. At the bottom is the spectrum of an 1868 comet called Brorsen. Above that is the spectrum of another 1868 comet called Winnecke II. And at the top is the spectrum of olive oil.

You can see that Comet Winnecke resembles olive oil more than it does Comet Brorsen. However, nobody deduced the existence of olive oil on the comets. (It would be an important discovery if it could be made.) But instead what this similarity shows is that a molecular fragment, diatomic carbon or C_2—two carbon atoms attached together—is present when you look at the spectrum of the comets and also when you look at natural gas and the vapor from heated olive oil. This is the discovery of an organic molecule, not one very familiar on Earth because of its instability when it collides with other molecules. It requires something close to a high vacuum, which does not naturally occur on the surface of the Earth. In the vicinity of a cometary coma, there is a high vacuum sufficient for C_2 not to be destroyed, and so here it is—the first discovery of an extraterrestrial organic molecule. And it turns out not to be one with which we have great familiarity.

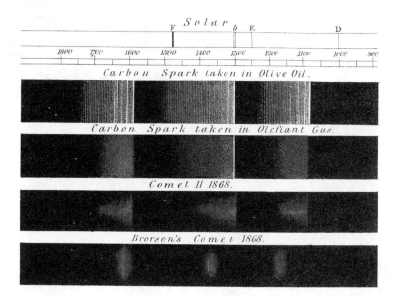

fig. 21

Spectrum of Comet 2001 Q4 (NEAT) on 2004 May 14

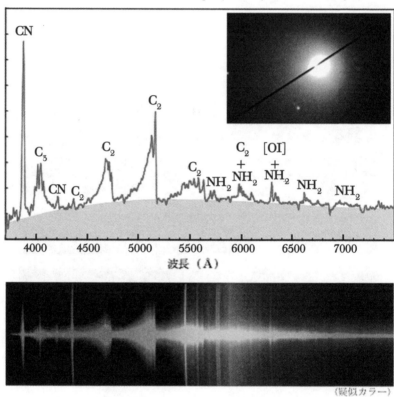

fig. 22

Here is a typical modern cometary spectrum, and we can see the prominent bands of C_2 and other things, too. We see NH_2, the amino group that is produced by dissociation of ammonia, NH_3, which is also the defining molecular group of the amino acids, the building blocks of proteins. And we see here the molecular fragment that caused all the trouble, CN, the nitrile or cyanide molecule.

A single grain of potassium cyanide on the tongue will instantly kill a human being. Discovering cyanide in comets worried people.

Especially when it appeared that in 1910 the Earth would pass through the tail of Halley's Comet. Astronomers tried to reassure people. They said it wasn't clear that the Earth would pass through the tail, and even if the Earth did pass through the tail, the density of CN molecules was so low that it would be perfectly all right. But nobody believed the astronomers. Perhaps the Earth did pass through the edge of the tail. In any case the comet came and went, nobody died, and in fact nobody could detect a single additional molecule of CN anywhere on the Earth. William Huggins, however, did die at the time that the comet came by, but not of cyanide poisoning.

Now, when we look closely at a comet, there is a tiny nucleus, the solid body that constitutes the comet everywhere except when it's very close to the Sun. The icy nucleus is typically a few kilometers across—but when it comes close to the Sun, the icy nucleus outgasses mainly water vapor and produces the coma and a long and lovely tail.

Consider the molecules we have just talked about: CN, C_2, C_3, NH_2. What are their parent molecules? Where did they come from? There are some precursors. We are seeing only fragments that have been chopped off of a bigger molecule by ultraviolet light from the Sun and the solar wind. It is clear that there is a repository of much more complex molecules—much more complex organic molecules—that are part of the cometary nucleus but which we have not yet discovered.

Radio astronomical studies have already found HCN (hydrogen cyanide) and CH_3CH (acetonitrile) in at least one comet. And these are interesting organic molecules that in other ways are implicated in the origin of life on Earth.

Imagine the air in front of your nose, highly magnified, say 10 million times. You would see a multitude of molecules, nitrogen and oxygen molecules, and occasional molecules of water

fig. 23

vapor and carbon dioxide. Air, as you know, is mainly oxygen and nitrogen. Now, if you take some air and cool it, you will progressively condense out the various molecules. Water will condense out first, carbon dioxide next, oxygen and nitrogen much later; that is, at much lower temperatures.

Let's consider the condensation of the water molecule. When condensation happens, it's not just that the water molecules drop out of the air helter-skelter. In fact they form a lovely hexagonal crystal lattice, which stretches off as far as the ice crystal or snowflake or whatever it is goes. Other molecules condense out at much higher temperatures, like silica, for example (silicon dioxide), which also forms a crystal lattice.

Let's go back to the solar nebula from which, as we said earlier, the solar system almost surely formed, with a protosun in the center and the temperature declining the farther we get from the Sun. Now we must imagine this as a mix of cosmically abundant materials, including water (H_2O, which we know through spectroscopic analysis of astronomical images is very abundant), methane (CH_4; we know that's very abundant), silica (SiO_2; we know that's very abundant), and what happens is that at different distances from the Sun, different materials will condense out, because they have different vapor pressures or different melting points. And what we see is (guess what?), water condenses out roughly at the vicinity of the Earth, whereas silicates condense out closer to the Sun, so liquid silicates or gaseous silicates are not to be expected under ordinary planetary conditions, even at the orbit of Mercury. Whereas you have to go out to somewhere near the present distance of Saturn before methane condenses. Now, methane is probably the chief carbon-containing molecule in the cosmos, and what this says is that in the early stages of the formation of the solar nebula there should have been a preferential condensation of methane in the outer parts of the solar system, but not in the inner parts. And if that is generally true, then we ought to expect more organic matter in the outer parts and much less in our neck of the cosmic woods.

Well, there is certainly not a huge amount of methane on the Moon or Mercury. But when we do go out to the orbit of Saturn we start finding not only evidence for methane—the planets Jupiter, Saturn, Uranus, and Neptune have lots of methane in their spectra—but we find a set of data that strongly implies the presence of complex organic molecules in the outer solar system.

This is a photograph of Iapetus, one of the outer moons of Saturn. The gray area is not in shadow. There is actually a remarkable division of one hemispheric surface into dark material and the other hemisphere into bright material. And the clear spectral signature of water ice is present in the bright areas.

We did not fly very close with either *Voyager 1* or *Voyager 2* to Iapetus. We think this is organic matter. It is very dark. At the center of this dark stuff, the albedo, the reflectivity, is something like 3 percent. I can't be sure, but I suspect that there is nothing in the room you are sitting in as dark as 3 percent albedo. Also, it is reddish. That is, it does not reflect much light, but it reflects more light in the red than in the blue part of the visible spectrum. And the values of the albedo and color are inconsistent with a wide range of other materials that you might offhand guess it might be—various of the salts, for example. They are very consistent with complex organic matter of various sorts. We know there is complex organic matter out there. I gave you one argument from the comets. Another argument is a category of meteorites called carbonaceous meteorites that fall to Earth, and they have several percent to as much as 10 percent of complex organic matter in them.

fig. 24

fig. 25

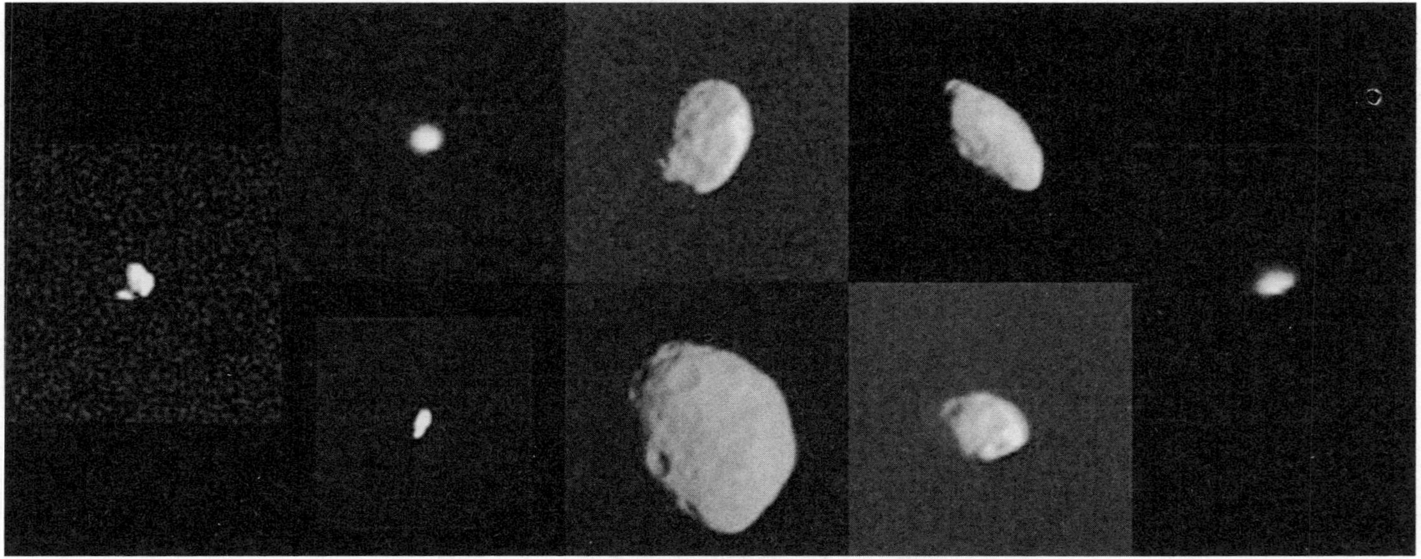

This is a family portrait of some of the small moons of Saturn. All of them were discovered by the *Voyager* spacecraft. None of these were known before. The smallest ones are maybe ten kilometers across. The biggest one may be a hundred kilometers. They're little worlds, and all of them are dark and red like Iapetus.

These are rings of Uranus. You may not think it's a very good picture, but it took an awful lot of work to make it. The picture was taken at 2.2 microns, in the infrared part of the spectrum. The rings are known to be quite different from the rings of Saturn. They are thinner, they are wispier, and they are black, again suggesting the prevalence of dark, reddish, presumably organic matter in the outer solar system.

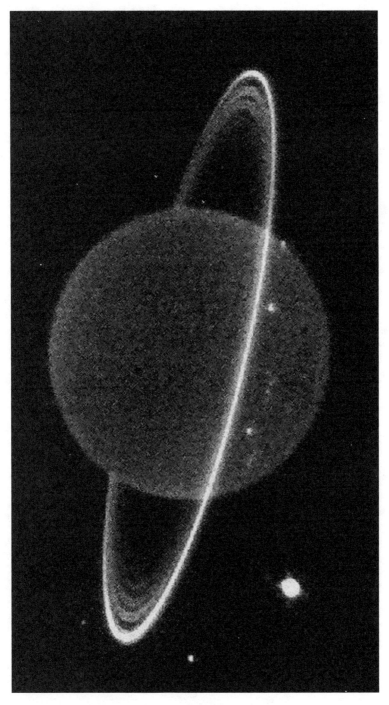

fig. 26

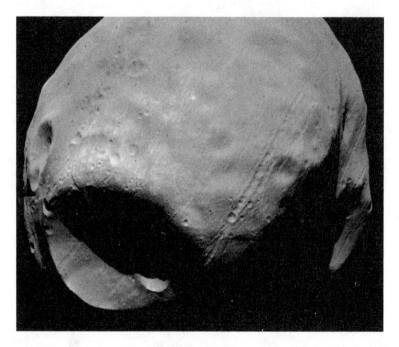

fig. 27

Now, this is not in the outer solar system. This is Phobos, the innermost moon of Mars, which may or may not be a captured asteroid from farther out in the solar system, and it too has this dark, reddish composition. Its mean density is known, and it is consistent with organic matter.

Deimos is the outermost Martian moon. Despite its different appearance from Phobos, it is likewise very dark, very red, same story.

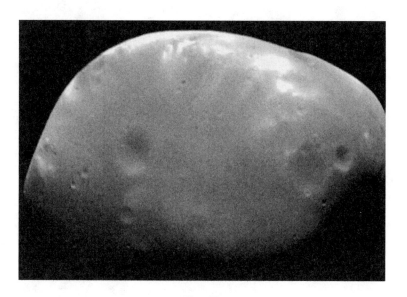

fig. 28

fig. 29

And I should mention that Mars itself, around which Phobos and Deimos are orbiting (all that rocky stuff is Mars, and the foreground instrumentation is the *Viking 1* Lander), at least in the two places that we landed with *Viking 1* and *Viking 2*, shows not a hint of organic matter. I will return to Martian exploration later, but I want to stress that the limits to the presence of organic matter on Mars are very low. There is not one part in a million of simple organic molecules and not one part in a billion of complex organic molecules. Mars is very dry, denuded in organic matter, and yet there are these two moons that may be made entirely of organic matter orbiting it. It's an interesting dilemma. These are two trenches that were dug by this sample arm in the Martian soil. So we gathered material from the subsurface and withdrew it back into the spacecraft and examined it with a gas chromatograph/mass spectrometer for organic matter, of which there was none.

I want to continue the story about organic matter in the outer solar system. And the best story by far, the one that we have the most information on, although it is still quite limited, is for Titan. Titan is the largest moon in the Saturn system. It is remarkable for many reasons, the most striking of which is that it is the only moon in the solar system with a significant atmosphere. The surface pressure on Titan (we know from *Voyager 1*) is about 1.6 bars, that is, about 1.6 times what it is in the room I am in as I write this. Since the acceleration due to gravity is about one-sixth on Titan what it is here on Earth, there is ten times more gas in the Titanian atmosphere than in the terrestrial atmosphere, which is a substantial atmosphere.

The organic molecules found in the gas phase in the atmosphere of Titan by the *Voyager 1* and *2* spacecraft include hydrogen cyanide (HCN, which we've talked about before), cyanoacetylene, butadiene, cyanogen (which is two CNs glued together), propylene, propane (which we know), acetylene, ethane, ethylene (these are all components of natural gas). Methane, likewise. And the principal constituent of the atmosphere, there as here, is molecular nitrogen.

It is, I think, very interesting that we have a world in the outer solar system that is loaded with the stuff of life. And we can calculate, at the present rate at which these materials are being formed on Titan, how much of this stuff has accumulated during the history of the solar system. The answer is the equivalent of a layer at least hundreds of meters thick all over Titan, and possibly kilometers thick. The difference depends on how long a wavelength of ultraviolet light can be used for such synthetic experiments. And, incidentally, there is also a range of entertaining evidence that there is a surface ocean of liquid

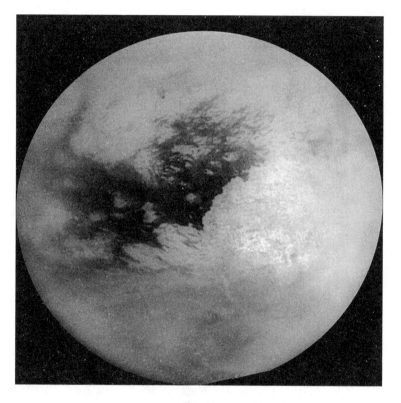

fig. 30

fig. 31

hydrocarbon at Titan.* So just think of that environment. There's land; probably there's ocean. The land is covered with this organic muck that falls from the skies. There is a submarine deposit underneath this ocean of liquid ethane and methane of more of this complex stuff, and then down deep is frozen methane and frozen water and so on.

Now, that's a world worth visiting. What's happened to that stuff over the last 4.6 billion years? How far along has it gotten? How complex are the molecules there? What happens when occasionally there is an external or an internal event that heats things locally and melts some ice and makes some liquid water? Titan is a world crying out for detailed exploration, and it seems to be a planetary-scale experiment on the early steps that here on Earth led to the origin of life but there on Titan were very likely frozen, literally, at the early stages because of the general unavailability of liquid water.

Likewise, there is a very stunning range of studies—mainly in the last two decades—of interstellar organic matter: not just a multiplicity of worlds in our solar system but the cold, dark spaces between the stars are also loaded with organic molecules.

---

*In July 2006, NASA announced that the Cassini space probe in the Saturn system observed evidence for numerous great lakes of liquid hydrocarbons on Titan.

We are looking toward the center of the galaxy in the direction of the constellation Sagittarius. You can see a set of dark clouds, some quite extensive, some much smaller. It is in these giant molecular clouds that well upwards of 50 different kinds of molecules have been found, most of which are organic. And it is precisely in such dark clouds that the collapse of solar nebulae is expected to happen, and therefore the forming solar systems should be composed, in part, of complex organic matter. The conclusion is that complex organic materials are everywhere.

Now let's return to the question of the origin of life on Earth. The organic stuff could have fallen in during the formation of the Earth, or it could have been generated in situ from simpler materials on the Earth in the same way as on Titan. At the present time there is no way of assessing the relative contributions from these two sources. What seems clear is that either source would be sufficient—adequate.

The Earth formed from the collapse of lumps of matter of the sort we talked about earlier, condensing from the solar nebula. Therefore in its final stages of formation, it was collecting objects that collided at high velocity and produced a set of catastrophic events, including the melting of much of the surface. This, it turns out, was not a good environment for the origin of life, as you might have suspected. But after a while, when the final sweeping up of the debris in the solar system was more or less completed, water delivered from the outside or outgassed from the inside started forming on the surface, filling in the ancient impact craters. And a trickle of material was still falling in from space. At the same time, electrical discharges and ultraviolet light from the Sun and other energy sources produced in-

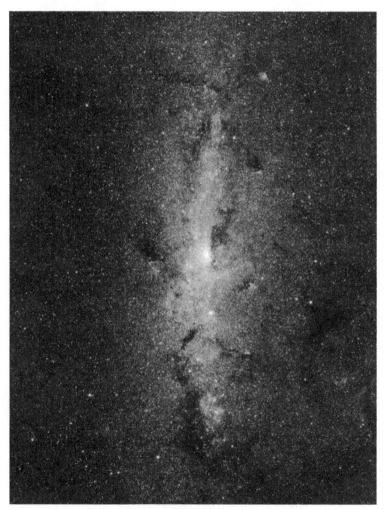

fig. 32

digenous organic matter. The amount of organic matter that could have been produced in the first few hundred million years of Earth history was sufficient to have produced in the present ocean a several-percent solution of organic matter. That is just about the dilution of Knorr's chicken soup, and not all that different from the composition either. And chicken soup is widely known to be good for life. In fact, it is just this warm, dilute soup, in the words of J. B. S. Haldane, who was one of the first two people to realize that this sequence of events was likely, in which the standard scenario for the origin of life occurs.

In the laboratory we can take molecules of water, ammonia, and methane—rather like the ones we've been talking about for Titan—and dissociate them by ultraviolet light. The fragments make a set of precursor molecules, including hydrogen cyanide, which then combine and, in water, form the amino acids. In such experiments not just the building blocks of the proteins but the building blocks of the nucleic acids are routinely produced. There is a range of subsequent experiments, in which the smaller molecular building blocks join together to form large and complex molecules.

If we look at the fossil record, we find that there is a range of evidence for microfossils dating back not just to the beginning of the Cambrian but dating back to as much as 3,500 million years ago.

Now, just think about these numbers. The Earth itself forms about 4,600 million years ago. Because of the final stages of accretion, we know that the Earth environment was not suitable for the origin of life back then. From studies of the late cratering on the Moon, it looks—since the Earth and the Moon were presumably in the same part of the solar system then as now—as if the Earth was not in a suitable state for the origin of life until perhaps 4,000 million years ago. So if the Earth is not appropriate to the origin of life until 4,000 million years ago and the first fossils are around 3,500 million years ago, then there are only about 500 million years for the origin of life. But those earliest fossils are by no means extremely simple organisms. They are, in fact, colonial algal stromatolites, and a great deal of evolution had to precede them. And that therefore says that the origin of life happened in significantly less than 500 million years. We don't know how much less. Six days was once a popular hypothesis. It's not excluded by these data, but at least it cannot be as long as 500 million years. It must have happened very fast. A process that happens quickly is a process that in some sense is likely. The faster it happens, the more likely it is. There is a difficulty in extrapolating from a single case; nevertheless this evidence suggests that the origin of life was in some sense easy, in some sense sitting in the laws of physics and chemistry. And if that's true, that is a very important fact for the consideration of extraterrestrial life.

There is a classic objection to this kind of argument about the origin of life. As far as I know, this objection was first posed by Pierre Lecompte du Noüy in a 1947 book called *Human Destiny* and is regularly rediscovered about once every half decade. It goes something like this: Consider some biological molecules. Not all of them. We'll give the evolutionists the benefit of the doubt. Let's just take a small, simple one, not something thousands of amino acids long. Let's pick an enzyme with a hundred amino acids. That's a very modest enzyme. Now, a way to think of it is as a kind of necklace on which there are a hundred beads. There are twenty different kinds of beads, any one of which could be in any one of these positions. To reproduce the molecule precisely, you have to put all the right beads—all the right amino acids—in the molecule in the right order. If you were blindfolded while assembling a necklace from equally abundant beads, the chance of getting the right bead in the first slot is 1 chance in 20. The chance of getting the right bead in the second slot is also 1 chance in 20, so the chance of getting the right bead in the first and second slots simultaneously is 1 chance in 20^2. Getting the first three correct is 1 chance in 20^3, and getting all hundred correct is 1 chance in 20^{100}. Well, you can see 20^{100} is 2^{100} x 10^{100}. And since 2^{10} is a thousand, which is 10^3, then 2^{100} is 10^{30}, so this is the same as 10^{130}. One chance in 10^{130} of assembling the right molecules the first time. Ten to the hundred-thirtieth power, or 1 followed by 130 zeros, is vastly more than the total number of elementary particles in the entire universe, which is only about ten to the eightieth (10^{80}).

So let's imagine that every star in the universe has a planetary system like ours. Let's say one planet has oceans. Let's suppose that the oceans are just as thick as ours. Let us suppose that there is a few-percent solution of organic matter in every one of those oceans and that in every tiny volume of the ocean that has enough molecules there is an experiment happening once every microsecond to construct this particular hundred-amino-acid-long protein. So in the ocean every microsecond an enormous number of these little experiments are going on. And identical things are happening in the next star system and the next star system, filling an entire galaxy. And then not just in that galaxy but in every galaxy in the universe. It turns out that if that sequence of experiments had gone on for the entire history of the universe, you could never produce one enzyme molecule of predetermined structure. And in fact it's much worse than that.

If you did that same experiment once every Planck time, the shortest unit of time that is permissible in physics, you still couldn't generate a single hemoglobin molecule, from which many people have decided that God exists, because how else do you make these molecules? If you haven't heard this before, doesn't this seem like a pretty compelling argument? Strong argument, right? A whole universe of experiments once every Planck time. Can't beat that.

Now let's take another look. Does it matter if I have a hemoglobin molecule here and I pull out this aspartic acid and I put in a glutamic? Does that make the molecule function less well? In most cases it doesn't. In most cases an enzyme has a so-called active site, which is generally about five amino acids long. And it's the active site that does the stuff. And the rest of the molecule is involved in folding and turning the molecule on or turning it off. And it's not a hundred places you have to explain, it's only five to get going. And 20^5 is an absurdly small number, only about 3 million. Those experiments are done in one ocean between now and next Tuesday. Now, remember what it is we're trying to do: We're not trying to make a human being from scratch, to have all the molecules of a human being fall simultaneously together in a primitive ocean and then have someone swim out of the water. That's not what we're asking for. What we're asking for is something that gets life going, so this enormously powerful sieve of Darwinian natural selection can start pulling out the natural experiments that work and encouraging them, and neglecting the cases that don't work.

So it turns out here, as in some of the arguments I was talking about yesterday, there is an important point that is left out in these apparent deductions of divine intervention by looking at the natural world. A very dramatic, strong statement of this sort has been made by the astronomers Fred Hoyle and N. C. Wickramasinghe. And their phrase, after a calculation in this spirit, goes something like this:

They say it is no more likely that the origin of life could occur spontaneously by molecular interaction in the primitive ocean than that a Boeing 747 would be spontaneously assembled when a whirlwind passed over a junkyard. That's a vivid image. It's also a very useful image, because, of course, the Boeing 747 did not spring full-blown into the world of aviation; it is the end product of a long evolutionary sequence, which, as you know, goes back to the DC-3 and so on until you get to the Wright biplane. Now, the Wright biplane does look as if it were spontaneously assembled by a whirlwind in a junkyard. And while I don't mean to criticize the brilliant achievement of the Wright brothers, as long as you remember that there is this evolutionary history, it's a lot easier to understand the origin of the first example.

I want to close on a beautiful little piece of poetry written by a woman in rural Arkansas. Her name is Lillie Emery, and she is not a professional poet, but she writes for herself and she has written to me. And one of her poems has the following lines in it:

> My kind didn't really slither out of a tidal pool, did we?
> God, I need to believe you created me:
> we are so small down here.

I think there is a very general truth that Lillie Emery expresses in this poem. I believe everyone on some level recognizes that feeling. And yet, and yet, if we are merely matter intricately assembled, is this really demeaning? If there's nothing in here but atoms, does that make us less or does that make matter more?

Four

EXTRATERRESTRIAL
INTELLIGENCE

There was a time when angels walked the Earth.
Now they cannot even be found in Heaven.

• Yiddish proverb •

If there is as a continuum from self-reproducing molecules, such as DNA, to microbes, and an evolutionary sequence continuum from microbes to humans, why should we imagine that continuum to stop at humans? Why should there be an open-ended gap in the spectrum of beings? And isn't it a little suspicious that the gap would begin with us?

It's of interest to me that our language has not really any appropriate terms for such beings. The theological languages have terms like angels and demigods and seraphim and so on. Even here it's interesting that the theological expectations of beings superior to humans generally represent a hierarchy of power but not of intelligence. And here again I think it is clear that we have imposed human values onto the universe. Certainly on this planet it is not apparent that there are beings more intelligent than humans, although a case can be made for dolphins and whales, and in fact if humans succeed in destroying themselves with nuclear weapons, a case could be made that *all* the other animals are smarter than humans.

I would like to describe a famous case of the search for extra-terrestrial intelligence—the search for beings more advanced than we—a case that failed. I want to explore why it failed, what lessons we can learn from this failure, and then move on to the modern search for extraterrestrial intelligence. I hope to stress where we have to be extremely careful, where we must demand the most stringent and rigorous standards of evidence precisely because we have profound emotional investments in the answer. Later I will attempt to use those skeptical stric-tures to apply more directly to the more conventional God hypothesis.

I suppose an equally good epigram for this subject is the fol-lowing sentence said by John Adams, second president of the United States, but long before he was that. As a lawyer and ad-vocate, he argued in defense of the British soldiers who were being tried at the Boston Massacre trials in December 1770. And he did this not because he was in favor of the British cause. He wasn't. He defended those he opposed because he believed that the truth should be pursued above all other considerations. He said, "Facts are stubborn things; and whatever may be our wishes, our inclinations, or the dictates of our passions, they cannot alter the state of facts and evidence." Well, sometimes they can, but we hope they can't.

The year is 1877, let us imagine. The motion of the Earth around the Sun and Mars around the Sun has brought Mars and the Earth close together, as they tend to be at intervals of roughly seventeen years.

An Italian astronomer named Giovanni Schiaparelli, look-ing through a newly completed and fairly large aperture tele-scope in Italy, was glancing at Mars and suddenly saw the surface of the planet reveal a profusion of intricate, fine, linear detail that a later observer described as being like the lines in a fine steel etching. Schiaparelli promptly called these lines *canali,* an Italian word meaning "channels" or "grooves." We can understand how it was translated into English as "canals," a word with a clear imputation of design, of intelligence, of vast engineering works constructed for a reason. The idea of *canali* on Mars was taken up by an American astronomer named Percival Lowell, a wealthy Bostonian. Lowell constructed a major observatory, with funds out of his own pocket, near Flagstaff, Arizona, called, naturally, the Lowell Observatory, to study these markings.

Lowell was convinced that Schiaparelli was right, that the planet was covered by a network of intersecting single and double straight lines, that these lines passed over enormous distances and therefore could correspond only to engineering works on the most massive imaginable scale. Other observers also found the canals; that is, drew them. Photographing them was much more difficult. The argument was that atmospheric "seeing" was unreliable, due to the intrinsic turbulence and unsteadiness of the Earth's atmosphere, which generally prevent you from seeing the canals. But every now and then, by chance, the atmosphere steadies, the turbulent eddies of air are not in your line of sight to Mars, and just for a moment you can see the planet as it truly is with this network of straight lines. And then another bit of atmospheric turbulence comes by and the planetary image becomes shimmery and the details are lost. Lowell reasoned that a photograph, which involves a time exposure that adds up the rare moments of good seeing with the much more plentiful moments of bad seeing, would not reveal the canals. But the human eye can remember those instants of excellent seeing and reject the other moments, much more common, when the image is fading and blurring and distorting. And this is why, he argued, experienced observers skilled in drawing what they see at the telescope could obtain results that the photographic emulsion could not.

There were other astronomers who, for the life of them, couldn't see the straight lines, but there was a range of explanations: They were not in the best sites for their telescopes. They were not experienced observers. They were not adequate draftsmen. They were biased against the idea of canals on Mars.

Lowell and Schiaparelli were by no means the only astronomers who could find the canals. Astronomers all over the world saw them, drew them, mapped them, named them. And there were literally hundreds of individual canals that were named.

There was a point of view that said that the canals were not really on Mars, that they represented some sophisticated failure of the human hand-eye-brain combination, that Lowell and his confreres were too carried away by the power of the idea. Lowell, who was a superb popular expositor, dismissed these objections in various ways and pointed to the remarkable similarity of the maps that he had drawn to those that other independent observers had drawn, say, for example, W. H. Wright at the Lick Observatory. Lowell argued that this convergence by quite separate observers, with no prior collusion, onto the same pattern of straight lines could only be due to something on Mars, not to something on the Earth. Lowell deduced from these straight lines an ancient civilization on Mars more advanced than we, having to face a planetary drought of proportions unprecedented on Earth. And their solution was to construct a vast, globe-girdling network of canals to carry liquid water from the melting polar caps to the thirsty inhabitants of the equatorial cities. What's more, it was possible to conclude, Lowell thought, something of the politics of the Martians, because the network crossed the entire planet. Therefore there was a world government on Mars, at least as far as engineering detail went. And Lowell went so far as to be able to identify the capital of Mars, a particular spot on the surface called Solis Lacus, the Lake of the Sun, from which six or eight different canals seemed to emanate.

Now, this is a lovely story. It passed into the popular consciousness, into folk literature, was most powerfully impressed on the global consciousness through H. G. Wells's *War of the Worlds*, through a set of science-fiction novels by Edgar Rice Burroughs (the man who invented Tarzan), and then in 1938 by Orson Welles's "War of the Worlds," broadcast in America on the eve of the Nazi invasion of Europe, at a time when fears of a distinctly terrestrial, not extraterrestrial, invasion were in everybody's mind.

And yet there are no canals on Mars. Not one. The whole thing is wrong. It's a mistake. It is a failure of the human hand-eye-brain combination. Lowell's idea evoked a passion, I think a very understandable and humane passion. The vision of more advanced beings on a neighboring planet, with a world government, struggling to keep themselves alive, was a wonderful idea. It was so wonderful that the wish to believe it trumped the scrupulousness of the investigative process.

So what can we conclude from this? Well, we can conclude that in a sense Lowell was right, that the canals of Mars are a sign of intelligent life. The only question is which side of the telescope the intelligent life is on. And as we see, the intelligent life was on our end of the telescope. People staked their careers on an observable phenomenon, apparently reproducible by others in quite different parts of the world. A huge public concern and interest were generated. This was only one of several different arguments for intelligent life on Mars today, all of which are mistaken.

If scientists can be fooled on the question of the simple interpretation of straightforward data of the sort that they are routinely obtaining from other kinds of astronomical objects, when the stakes are high, when the emotional predispositions are working, what must be the situation where the evidence is much weaker, where the will to believe is much greater, where the skeptical scientific tradition has hardly made a toehold— namely, in the area of religion?

Let's think about the question of extraterrestrial intelligence. There are several approaches. There is one that says, well, it is a vast universe. There must be beings much smarter than we are. They must have capabilities vastly in excess of ours. Therefore they should be able to come here. If we are poking around in neighboring worlds in our planetary system, then should not intelligent beings elsewhere in our solar system, as Lowell thought, or in other planetary systems, of which we now know there are many, shouldn't they be visiting here? And that then takes us to the issue of unidentified flying objects and ancient astronauts, which we will get to. But here I would like to concentrate on what is now the mainstream scientific approach to the issue of extraterrestrial intelligence, one that I should say from the beginning I have been deeply involved in and support wholeheartedly. But at the same time I think it sheds light on this question of what is suitable evidence and what isn't.

At what moment do you say that the evidence is sufficient to deduce the presence of extraterrestrial intelligence? I believe that while the details are slightly different, the argument is not significantly different from the question, what would be convincing evidence of the existence of an angel or a demigod or a god? First off, there's the question, is it plausible? That is, whatever you do to search for extraterrestrial intelligence, it is going to cost some money. You want a plausibility argument first that it makes at least a little sense. Clearly, were we to find extraterrestrial intelligence, this would be a discovery of enormous importance scientifically, philosophically, and, I maintain, theologically. But you'd want to have some expectation of success, some argument to counter skeptics who might say, "There is no evidence that we have been visited; therefore it is a waste of time."

So what we would really like to know is how many sites of intelligent beings, more intelligent than we, there are in, say, the Milky Way Galaxy? And how far is it from here to the nearest one? If it turns out that the nearest one is some immense distance away—let's say, at the center of the Milky Way Galaxy, 30,000 light-years—then we might conclude that the prospects of contact are small. On the other hand, if it turns out that the nearest such civilization is relatively nearby—let's say, a few tens or even a few hundreds of light-years—then it might make sense in some way, which I'll go into, to try to search for it.

Now, a convenient approach to this issue (it is hardly precise) is what is called the Drake equation, after the astronomer Frank Drake, who has been a pioneer in the scientific approach to this question. And it goes roughly like this: There is a number, call it N, of technical civilizations in the Galaxy, civilizations with the technology to permit interstellar contact (that technology essentially is radio astronomy). That number is

$$N = R \times f_p \times n_p \times f_l \times f_i \times f_c \times L$$

the product of a set of factors, each of which I will define. (All that is involved in this equation is the idea that a collective probability is the product of the individual probabilities, quite like what we were talking about earlier on the probability that the right amino acid is in the first slot in the protein, and in the second slot, and in the third slot, and then you multiply those probabilities. The chance that you'll get heads in the first coin toss is one-half, the chance that you'll get heads in the second toss is one-half, the chance that you will get two consecutive heads is a quarter, three consecutive heads is an eighth, and so on.)

So the number of such civilizations depends on the rate of star formation, which we call R. The more stars that are formed, the more potential abodes for life there will be if they have planetary systems. That seems clear. Multiply that figure times f_p, the fraction of stars that have planetary systems. But it's not good enough just to have planets; they have to be suitable for life. So multiply by n_p, the number of planets in an average system that are ecologically suitable for the origin of life, then times f_l, the fraction of such worlds in which life actually arises, times f_i, the fraction of such worlds in which over their lifetime intelligent life evolves, times f_c, the fraction of such worlds in which the intelligent life develops a technical communicative capability, times L, the lifetimes of the technical civilization, because clearly if civilizations destroy themselves as soon as they are formed, everything else may go swimmingly well and yet there would be nobody for us to talk to.

So let me give my wild guesses about what these numbers are. I stress that we don't know these numbers very well, that our uncertainty progressively increases as we go from the leftmost to the rightmost factor. And that the largest uncertainty by far is in L, the lifetime of a technical civilization.

There are some hundred thousand million stars in the Milky Way Galaxy.

The lifetime of the Milky Way Galaxy is something like ten

thousand million years, and therefore a modest average estimate of the rate of star formation is about ten stars per year. A very interesting number, that, by itself. Every year there are ten new suns that are born in the Milky Way Galaxy, and many of them, probably, with planetary systems. And billions of years from now, maybe they will have life.

On the question of the fraction of stars that have planets going around them, I previously talked about the burgeoning recent evidence from ground-based and space-based observatories for planetary systems, both those just forming and ones that are fully formed around nearby stars. The statistics are remarkable. The IRAS satellite data alone suggest that something like a quarter of nearby main sequence stars a little younger than the Sun have something like a solar nebula in the process of formation. It's an amazingly large number. And any of them that have fully formed planetary systems we can detect only in certain special cases. You would not expect that every star has a planetary system, but the number looks very large. Just for the sake of argument, I'll take the fraction f_p to be something like a half. Now consider the number of planets per system that are in principle suitable for the origin of life. Well, certainly in our system, we know at least one, the Earth. And good arguments can be made that it is possible on other planets, on other bodies. We talked about Titan. There is an argument for Mars. Not to pretend any kind of accuracy, but just so that we can put in numbers that easily multiply each other, let us take that number, n_p, as two.

The fraction of ecologically suitable planets in which life actually does arise over a period of hundreds of millions or thousands of millions of years, I will take to be very high, on the basis of the sorts of arguments I made earlier, especially the speed with which the origin of life seemed to have happened on this planet. So I'll take f_l to be around one.

And now we come to more difficult numbers. Life has arisen on a given planet, and you have thousands of millions of years in which the environment is somewhat stable. How likely is it that intelligence and technical civilizations arise? On the one hand, we might argue that there are a sequence of individually unlikely events that must happen for humans to evolve. For example, the dinosaurs had to be extinguished, because they were the dominant organisms on the planet and our ancestors in the times of the dinosaurs were furry, scurrying, burrowing creatures, about the size of mice. And it is only because of the extinction of the dinosaurs that our ancestors could get going. And the extinction of the dinosaurs seems to have been caused by an immense collision by an asteroid or cometary nucleus with the Earth some 65 million years ago at the end of the Cretaceous period. That is a statistical event, and if that had not happened, maybe I would be ten feet tall with green scales and sharp, pointy teeth, and you would be similarly tall and green and pointy-toothed. We would both likely consider ourselves extremely attractive. What handsome fellows we are. And how strange it would seem if I proposed that had things gone differently, then the little mice that bother us might have evolved to become the dominant organism, and the only remnants of us would be salamanders and crocodiles and birds. That's on the one hand.

On the other hand, there is no reason to think that there is only one path to intelligent life. The selective advantage of intelligence is clearly high. Other things being equal, if you can figure the world out, you have a better chance of survival. At least until the invention of nuclear weapons.

Human brains comprise a significant fraction of our body

mass, more than for almost any other animal on the planet. And this then suggests a progressive development of brains to figure out the world. The more data processing, the more chances for survival we have. There is no reason to think that this is a peculiarly human situation, and it ought to be true on other planets as well.

So then we come to the question, if you have intelligent life, is it guaranteed to develop technical civilizations? Clearly not. The dolphins and whales are intelligent, based on many different anecdotal accounts and on the argument of brain-mass-to-body-mass ratio, and yet they have built nothing, because they don't have hands and they live in a different environment than we do.

It is easy to imagine a world full of poets who do not build radio telescopes. They're very smart, but we don't hear from them. So not every intelligent life-form need be technological or communicative. What this product of $f_i \times f_c$ is, no one really knows. We can certainly point out that it took most of the history of the Earth before Ornithoides or Cetacea or primates developed. They all developed in the last few tens of millions of years. Why did it take so long? Well, there's probably a certain degree of complexity that is essential for being able to figure things out.

On the other hand, the Earth and the solar system have thousands of millions of years more ahead of them, as do other planets as well. A number for $f_i \times f_c$ that I believe to be modest is 1/100—1 percent. (I do not at all say that I know what these numbers are; these are merely rough estimates to collect the various uncertainties together. I do not claim this is holy writ.) If we multiply these numbers together, $10 \times 1/2 \times 2 \times 1 \times 1/100$, the product is a tenth. So the number N of technical civilizations in our galaxy would be one-tenth times their average lifetime L in years. (L is in years because R was ten stars per year, and the product must not have any years in it, just the number of civilizations.)

So what is L? What is the lifetime of a technical civilization? We have had radio telescopes for only the last few decades. An argument could be made by reading the daily newspapers, among other things, that our civilization is in great peril. And therefore that, for the Earth at least, the lifetime of a technical civilization in this sense is a decade or a few decades. And if that number were typical for civilizations in general, then L would be, let's say, a decade, ten years. So let's call this the most pessimistic route. A tenth times ten is one, and the number of technical civilizations in the galaxy would be one. Where is it? It's us.

So there's nobody to talk to except ourselves, and we hardly do that very well. In that case, if you believe that argument, it would be foolish to make a massive or expensive search for extraterrestrial intelligence because even if this number L were a few decades, the number of civilizations would be only a few, and therefore the distance to the nearest one would be enormously far away.

Now let's take another route, the optimistic one. And that is, it seems perfectly possible that we are able to solve the issues of technological adolescence that confront us. And even if there's only a small chance of doing that, say, 1 percent, 1 percent of all those civilizations in the Galaxy living for very long periods of time implies a very large number. Suppose that 1 percent of civilizations lived for evolutionary or geological or stellar evolutionary timescales—say, billions of years. If there's only 1 percent that do that, then the average lifetime would be 1 percent of 10^9 which is 10^7, so 10 million years would be the value for L. Multiply that by a tenth and the answer would be a million, 10^6 civilizations in the galaxy, a vastly different story.

So you can see that while there are significant uncertainties for each of these factors, by far the largest uncertainty, the place we have the least experience (none whatever, as a matter of fact) is in the average lifetime of a technical civilization. And it is this connection of L with the number of civilizations and the distance to the nearest one that in a remarkable way binds up this quite outré subject of extraterrestrial intelligence with the most pressing human concerns. Because it means that the receipt of a message, never mind being able to decode it, from elsewhere would say that L is probably a large number, that someone has been able to survive technological adolescence. It would be knowledge very much worth having.

If there are a million technical civilizations in the Galaxy, then you can readily calculate to first order, just extracting a cube root, the distance to the nearest civilization. If they are randomly distributed through the Galaxy, and we know how many stars there are in a galaxy, how far to the nearest one? And the answer is, it's just a few hundred light-years away. It's next door. It's not next door as far as visiting, but it's next door as far as radio communication.

Now, even a few hundred light-years away means that we must not imagine much in the way of dialogue. It's more monologue. They talk, we listen, because otherwise they would say, let us imagine, "Hello. How are you?" and we would say, "Fine, thank you, and you?" and that exchange would take, say, six hundred years. It's not what you might call a snappy conversation.

On the other hand, it's very clear that one-way transmission of information is something that can be enormously valuable. Aristotle talks to us. We do not, except for spiritualists, talk to Aristotle. And I have grave doubts about the spiritualists. (Although Aristotle is almost never on their list of contacts.)

Now, let's therefore say a few more words about this idea of radio communication. What we imagine is that beings on a planet of another star know that emerging civilizations will stumble upon radio. It's part of the electromagnetic spectrum; it is, as I will show you in a moment, a clear channel through the Galaxy. The technology is relatively simple and inexpensive. Radio waves travel at the speed of light, faster than which nothing can go, so far as we know. The information that can be transmitted is enormous, not just "Hello, how are you?" Put another way, if an identical system were at the center of the Galaxy and we were here using our present detection technology, we could pick up that signal coming from thousands of light-years away. It gives you an idea of the remarkable power of this technology, which has in fact been only lately brought up to its actual capabilities.

There is a question of frequency. What channel would you listen on? There's an enormous number of possible radio frequencies. We have here the radio frequency spectrum in gigahertz, thousands of millions of cycles per second, against the noise background from various sources in degrees absolute. And what we see is that at the low frequencies there is a background from charged particles in magnetic fields in the Galaxy, the galactic background. It's noise. And it gets to be very substantial noise.

This is not where you would want to transmit or receive. At the high-frequency end, there is another source of noise, intrinsic to the quantum nature of radio detectors. And in the middle there is a broad region where the noise is low, and this is the window in which it makes sense to transmit. In this window there are certain spectral lines, for example, of atomic hydrogen, the most abundant atom in the universe, at specific frequencies. So for this reason there is now a very sophisticated search program going on at Harvard, in Massachusetts, a cooperative project with Harvard University and the Planetary Society, a hundred-thousand-member worldwide organization, and it is remarkable that dues and contributions to a private organization are able to maintain by far the most sophisticated search for extraterrestrial intelligence yet attempted.*

fig. 33

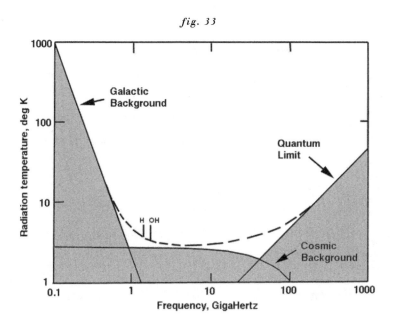

*In 2006 the Planetary Society and Harvard University inaugurated the SETI Optical Telescope, the first-ever optical observatory dedicated to the search for intelligent extraterrestrial signals. For the history of the Planetary Society and SETI, see www.planetary.org, and for the thrill of actually participating in the search, go to www.setiathome.ssl.berkeley.edu/.

This illustration might convey a sense of how a success would be noted. The slanting line indicates a very weak signal from an extraterrestrial source. You listen at many frequencies for a while and see if there's anything happening. The Planetary Society system was recently upgraded, so that 8.4 million separate channels are being listened to simultaneously. The antenna points to some part of the sky. And some places have peaks. They may be due to radio interference on the Earth, satellites in Earth orbit, automobile ignitions, diathermy machines. But each of those has a particular kind of signature, and it is possible to imagine signals that don't look like any of those things, which the computer immediately would cull out of the noise, leaving no doubt that this was an artificial signal of extraterrestrial origin, even if we had no opportunity, no ability, to understand what it meant.

Now, as I said, the expectation is that they send and we, newly emerged, the youngest communicative civilization in the Galaxy, we listen. Not the other way around.

Let me stress that this is the one respect in which our civilization is probably unique in the Galaxy. No one even slightly more ignorant can communicate at all. Let me say this in a better way: A civilization only a few decades behind us would not have radio astronomy and therefore could not tumble to this technique. Or maybe they could tumble to it, but they couldn't manifest it. And anyone, therefore, whom we hear from is likely to be ahead of us, because if they're even a little bit behind us, they can't communicate at all.

So the most likely situation is communications from beings vastly more advanced than us. And this therefore raises the ques-

fig. 34

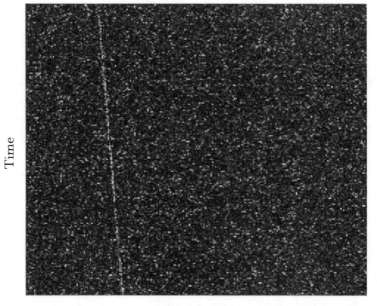

Time

Frequency

tion, could we possibly understand what they're saying? What we have to remember here is that if this is an intentional message from them to us, then they can make it easy. They can make allowances for civilizations. And if they do not choose to do that, then we will not understand the message.

Maybe you would say advanced civilizations communicate with each other by zeta waves. And I'd say, "What is a zeta wave?" And you reply, "It is something fantastic for communication that I can't give you any details about, because it won't be invented for another five thousand years." Well, that's wonderful, and if those fellows can communicate with zeta waves, that's terrific. But if they wish to communicate with us, they will have to wheel out some ancient, creaking radio telescope from the technology museum and use it, because that is all that young civilizations will be able to understand and detect.

Now, suppose we get a message. What would it be like? Here is a possibility: There would be a powerful beacon or announcement signal, something that makes it very clear that we are unambiguously receiving a message from an advanced civilization. It might, for example, be highly monochromatic; that is, a very narrow radio frequency band pass, and/or it might be a sequence of pulses that could not possibly be of natural origin. For example, a sequence of prime numbers, numbers divisible only by 1 and themselves—1, 2, 3, 5, 7, 11, 13, 17, 19, and so on. There is no natural process that could produce such numbers.

Then, having established unambiguously that the message was from intelligent beings in space, it is perfectly possible to imagine a vast amount of additional information conveyed in ways that we can understand. For example, it is perfectly possible to transmit pictures. In fact, it's done by radio all the time. That's what your television set does. It is possible to send mathematics. It's very easy. I mean, suppose they set out the numbers—

beep, that's one; *beep beep*, that's two; *beep beep beep*, that's three—and so on. And then they do (I'm just going to make this up) *beep glagga beep wonk beep beep*. Well, a few more like that and you decide a *glagga* means "plus" and *wonk* means "equal." But suppose they now do *beep glagga beep beep wonk beep beep?* And then there's some symbol after that. That symbol, that new symbol must mean "false." And you can immediately see that abstract concepts like true and false could be communicated very quickly. And between these two modes—the use of mathematics, which we would, of course, share in common, and the transmission of pictures—it is possible that a very rich message could be conveyed. What that message would be, clearly none of us are in a position to say.

Now, I would like you to just think about contrasting this open-minded, experimental approach, which consists of some plausibility arguments that no one takes too seriously, with the more traditional approach to intelligent life in space: the one in which there are no experiments, in which there is no withholding of opinion until the evidence is in, in which we are asked merely to take it on faith. The contrast is, in my opinion, very stark. There is quite a different approach in method. And I remind you about how powerfully we were fooled by the Martian canal situation, where passions and emotions were heavily engaged.

What do they look like? There is a standard Hollywood convention that extraterrestrials look just like us. Well, maybe they have pointy ears or antennae or green skin, but those are minor cosmetic variations. Extraterrestrials and humans are fundamentally the same. Why should that be? Look at the long sequence of stochastic random events that led to our evolution. I mentioned the extinction of the dinosaurs. That's one. Take another: We have ten fingers. And that's why we use base-ten arithmetic. Nothing special about one, two, three, four, five, six, seven, eight, nine, and then one-zero, except we count on our fingers. Why do we have ten fingers? Because we evolved from a Devonian fish that had ten phalanges in its fins. If we evolved from a Devonian fish that had twelve phalanges, then we'd all be doing base-twelve arithmetic, and base-ten arithmetic would be considered only by mathematicians.

This is true at every level, including biochemical levels, to such an extent that I think it is fair to say—never mind some other planet—if you started the Earth out again and let just these random factors operate, such as when a cosmic ray would strike a chromosome, producing a mutation in the hereditary material, you might wind up with intelligent beings after some thousands of millions of years. You might wind up with creatures of high ethical and artistic or theological accomplishment. But they would not look anything like human beings. We are the products of a unique evolutionary sequence. Unique doesn't mean better; it just means unique. Elsewhere, different environment, different necessity to adapt to changing conditions, a different sequence of random events, including random genetic events, and we should not expect anything like a human being.

Now, what about religion? What about the idea that we are all made in God's image? Is that also a failure of the imagination? What do we mean when we say we are made in God's image? Do we, for example, imagine that God has nostrils and breathes? If so, what does He breathe? Air? Where is the air? Air with oxygen in it? No other planet in the solar system has oxygen except the Earth. Why restrict God to very few places? Why would He need nostrils? What about a navel? Would God have a navel? What about hair? What about a vermiform appendix? What about toes? Toes are clearly the result of our ancestors' life in the canopy of the high forest, swinging from branch to branch. Very good to have four limbs that can hold on to trees. We just happened to have the toes in this particular transitional moment. Big toe is good for balance; little toe is not good for very much at all. It's just an evolutionary accident. Vermiform appendix? Likewise good for nothing. It's just on its way out.

Arthur Clarke has said that Christian orthodoxy is too narrow and timid for what is likely to be found in the search for extraterrestrial intelligence. He has said that the doctrine of man made in the image of God is ticking like a time bomb at Christianity's base, set to explode if other intelligent creatures are discovered. I don't in the least agree. I think that the only sense that can be put on the phrase "made in God's image" is that there is a sense of intellectual affinity between us and higher organisms, if such there be.

The same laws of physics apply everywhere. If we imagine those extraterrestrial beings sending us radio messages, we and they have something in common. We must. The very act of receiving the message means that we have radio technology in common. We have quantum mechanics. We have atomic physics. We have Newtonian gravitation. We can see that those laws of nature apply everywhere in the universe. It's not a question of what your biology is like. It's not a question of the sequence of events that led to you getting a technical civilization. The mere fact that you have a technical civilization means that you have come to grips to some extent with the universe as it really is. And so it is in that sense and in that sense alone, I believe, that it makes sense to talk about such an affinity between advanced beings and ourselves.

Five

EXTRATERRESTRIAL FOLKLORE: IMPLICATIONS FOR THE EVOLUTION OF RELIGION

I consider the idea of extraterrestrial intelligence a subject of philosophical, scientific, and even historical importance. If we were so lucky as to receive some sign of extraterrestrial intelligence, I think there is little doubt that it would be an extremely significant historical event. And if, on the other hand, we were to make a detailed and comprehensive search to no avail, that would also be something worth knowing. It would say something about the rarity and preciousness of intelligent life and again, I believe, would have extremely important and beneficial social consequences. Therefore the search for extraterrestrial life is one of those few circumstances where both a success and a failure would be a success by all standards.

So I am hardly opposed to the idea of extraterrestrials visiting us. If we ourselves are poking around our solar system, if we are capable, as we are, of sending our own spacecraft not just to the other planets in our solar system but beyond our solar system to the stars, then surely other civilizations, if they exist, thousands or millions of years more advanced than ours, ought to be able to achieve interstellar spaceflight much more readily, much more swiftly.

And I don't for a moment deny this as a possibility. I would stress that the economy of effort is far greater for radio communication than direct communication by interstellar spacecraft. I would argue that you can broadcast to millions or thousands of millions of worlds simultaneously, speedily, inexpensively, in a way that even for a very advanced civilization would be much more difficult and costly to do via interstellar spacecraft. However, I certainly could not exclude the possibility that the Earth is now or once was visited. But precisely because the stakes in the answer are high, precisely because this is an issue that engages powerful emotions, we would in this case demand only the most scrupulous standards of evidence.

I want tonight to discuss two modern hypotheses that I think are proper to call folklore, the ancient astronaut hypothesis and the UFO or unidentified flying object hypothesis, and then attempt to connect them with the history of slightly more conventional religions.

The ancient astronaut hypothesis was popularized most effectively by a Swiss hotelier named Erich von Däniken. And his works, the first of which was called *Chariots of the Gods?* (the question mark becoming suppressed in subsequent printings), were huge bestsellers in the late 1960s, early 1970s, selling worldwide tens of millions of copies, an enormously successful set of books.

The fundamental hypothesis of von Däniken was that there is impressed in the archaeology and folklore and myth of many civilizations on Earth certain indications of past contact with the Earth by extraterrestrial beings. This is not an absurd proposition on the face of it, but how acceptable the hypothesis is depends on how good the evidence is. And, unfortunately, the standards of evidence were extremely poor, in many cases non-existent. So to give an example (and I promise I am not bur-lesquing the argument as I describe it), here is von Däniken's approach to the pyramids of Egypt: The pyramids of Egypt, he said, are constructed of individual blocks, rectangular paral-lelopipeds, each of which weighs twenty tons or thereabouts. "Twenty tons," he said. That's extremely heavy. Individual per-sons could not lift a twenty-ton block, much less many of them, to make a pyramid. Therefore modern construction equipment is necessary, and in 2000 to 3000 B.C., that could only be of extra-terrestrial manufacture. Hence extraterrestrials exist.

Now, we can recognize that this argument neglects certain facts. If we knew nothing of Egyptian archaeology, we could nevertheless imagine ways in which large numbers of people could build massive edifices. (The Bible, after all, refers to am-bitious construction projects, for example the enormous Tower of Babel.) And then when we look at the internal evidence, or even read Herodotus, who alluded to Egyptian pyramid-construction techniques, we find that there is an entirely self-consistent and perfectly natural explanation. In fact, there are many, some of which involve sending rafts up the Nile, and rollers to move the blocks, and the removal of underlying material. There were even inscriptions on a few key blocks that say the equivalent of "My goodness, we did it!" signed "Tiger Team Eleven," which seems an unlikely delight in modest construction by some being who had effortlessly traveled through interstellar space. And we know that the first pyramid that was ever constructed fell down and that the second pyramid, halfway through construction, had the angle of the sides dramatically pared, because they had learned from the example of the first one that fell down. Again, an error of exceeding the angle of repose was unlikely to be made by an extraterrestrial spacefaring civilization.

Von Däniken noted that in Peru, in the plains of Nazca, there are large drawings on the desert that can properly be seen only from a great altitude. And they depict nothing very extraordinary in themselves: turkeys, condors, and other natural beasts and vegetables. But von Däniken wonders why anyone would construct something that could be seen only from a great altitude, from which he deduces not only that there were beings at a great altitude to see it but that such beings directed the construction, saying, "A little to the left." Now, in American football games, it is customary for people to be outfitted each with a square of cardboard on which is the fragment of a line or a letter. And at the appropriate moment, everybody holds up their piece, and from a great distance some symbol generally having to do with the hope for success of the home team is displayed. And yet no one deduces extraterrestrial intervention in such a case.

Or, von Däniken noted that in the Pacific, on Easter Island, there is a set of massive stone monoliths all facing to the sea, all of which are much too heavy to be lifted by one or two people and all of which, as Jacob Bronowski mentioned, look exactly like Benito Mussolini. They were quarried some substantial distance away on this very small island. And again von Däniken deduces extraterrestrial manufacture from the fact that he cannot himself figure out how people living before the industrial revolution could cut, transport, and erect these monoliths. And yet years before von Däniken wrote, Thor Heyerdahl had gone to Easter Island and with a small team using only the simplest of tools, had transported and erected one of these monoliths that had been found in a supine position. And the erection method included just shoving small bits of dirt and stone under one side of it until it got to the high, steeper angle and then finally stood up.

So there are many other such arguments by von Däniken, most of which have lower plausibility than the arguments I've just presented to you. I've presented some of his best cases. Fundamentally, what von Däniken has done is to sell our ancestors short, to assume that people who lived a few thousand years ago or even a few hundred years ago were simply too stupid to figure anything out, certainly to work together for a long period of time to construct something of monumental dimensions. And yet people of a few hundred or a few thousand years ago were no less intelligent than we are, no less able. Perhaps in some ways they were better able to work together. The argument is absurdly specious. So how do we understand that so specious an argument could have been so wildly successful (although today one does not hear much about ancient astronauts)? It's an interesting question.

I think the answer is absolutely clear. The emotional appeal of von Däniken made perfect sense. It was the hope that extraterrestrials would come and save us from ourselves. The hope that if they had intervened many times in human history, surely in the present time, a time of great crisis recognized in the 1960s and '70s and manifestly clear today in an age of fifty-five thousand nuclear weapons, that the extraterrestrials would come and prevent us from doing the worst to ourselves. And in that sense I consider it an extremely dangerous doctrine, because the more likely we are to assume that the solution comes from the outside, the less likely we are to solve our problems ourselves.

But ancient astronauts are only a sideshow, a minor codicil of the principal doctrine along these lines of the twentieth century, and that is flying saucers or unidentified flying objects. And here we have not just the writings of a half dozen people but some collective enterprise involving an enormous number of people all over the world, and something like 1 million separate sightings since 1947, when the term "flying saucer" was first coined.

The standard mythos is quite straightforward. It's that a device of exotic design and manufacture is seen in the sky, at least sometimes doing things that no machine of terrestrial manufacture could do. More rarely, it discharges exotic beings, who engage in conversations with terrestrials, capturing people from the Earth, performing exotic medical examinations on them, taking them to other planets, and occasionally having sexual congress with them, resulting in offspring who are fully human—a feat somewhat less likely, if we bear in mind the clear evidence of Darwinian evolution, than the successful mating between a man and a petunia.

Now, what would we require, if we took even a modestly skeptical approach, to be convinced? We would not require a million cases. I don't think we would require more than one, provided that one case were absolutely solid. We would require that that solid case be simultaneously very reliably reported and very exotic. It is insufficient to have several hundred people see it independently as a light in the sky. A light in the sky can be anything. It has to be much more concrete, much more specific. On the other hand, it is also insufficient to have, let us say, a twenty-meter-diameter, saucer-shaped, metallic object land in a suburban garden on Long Island, a seamless door open (there is some fascination with seamless doors in these stories), a four-meter-high robot walk out, pet the cat, pick a flower, wave to the startled householder, and then disappear back into the seamless door, which closes, and the craft takes off. If only one person saw it, the cat being unavailable for corroboratory testimony, then this likewise is not a compelling case. We would require that the examples be, simultaneously, extremely reliably reported and extremely exotic.

I have spent, although not recently, a great deal of time on UFO cases, feeling that it was my responsibility, since I'm interested in extraterrestrial life, to see if the problem has not been finessed, if the extraterrestrials are not here, in which case, of course, my colleagues and I would be saved a great deal of effort. I spent time on a committee established by the U.S. Air Force to look into this story and have interviewed some of the participants in a few of the most famous cases. And let me give you my overall impressions.

By no means are all UFO cases identified, established as to what they are. Some of them are too sparsely and scantily reported, and a few are sufficiently mysterious, so of course you couldn't expect that to be the case.

But let me give you a sense of routine UFO reports that have been checked out and we do know what they are:

The Moon. You may think that there is no way that someone could identify the Moon as an extraterrestrial spacecraft. But there are many cases where not only has that been done but the Moon has been reported as following and even harassing the observer.

The aurora borealis; bright stars; bright planets, especially under unconventional meteorological conditions; flights of luminescent insects; a low overcast, an automobile going up a hill, the headlights moving rapidly across the overcast; weather balloons.

There was a famous case in which a firefly was trapped between two adjacent panes of glass in an airplane cockpit window and the pilots were radioing about fantastic right-angle turns, defying the laws of inertia, estimated fantastic speeds.

They imagined it at some huge distance away instead of right in front of their noses.

Noctilucent and lenticular clouds, lens-shaped clouds, conventional aircraft with unconventional lighting. Unconventional aircraft.

Then there is a vast category of hoaxes. As soon as you could get your name in the newspaper by reporting a UFO, a lot more people started seeing UFOs than had done so previously. And some of them were done in good fun, some not. A famous case was a set of plastic bags from dry cleaners that were fashioned to form a hood around candles and the whole business sent aloft to make a small hot-air balloon, which can be done. And this very primitive technology was reported by hundreds of people as UFOs and performing maneuvers that, it was claimed, could not possibly have been performed. So there's a hoax plus some misapprehensions or flawed reporting, and the net result is something extraordinarily exotic. But it was only strange moving lights. This is one of the reasons I say that merely moving lights are insufficient.

Then there are cases of so-called high evidence. Photographs, for example. One of the earliest photographs of UFOs from the late 1940s was from a man named George Adamski, who was a space enthusiast and, in fact, identified himself in his first book as George Adamski of Mount Palomar. Mount Palomar was then the site of the largest optical telescope on the planet. And George Adamski was from Mount Palomar. He owned a hamburger stand at the base of Mount Palomar, in which he had a small telescope, and through that telescope he photographed wonders that the astronomers, consigned to the lofty recesses of the mountain, never saw.

One of his most famous photographs shows a clearly metallic, saucer-shaped object with three large spheres at the bottom, which he identified as landing gear and which later turned out to be a chicken brooder suspended by thread. This is one of those devices that encourages the eggs to hatch, and ordinary lightbulbs are used to warm it. And indeed there developed an entire detective industry to determine what common object was being photographed close up to explain this particular unidentified flying object case.

Now, I've probably made the point implicitly, but let me make it explicitly. I do not think there is any fundamental difference between this sort of UFO hoaxmongering and the sale of relics in the Middle Ages—pieces of the true cross and so on. The motivations are almost identical.

There are also cases, and Adamski was one of them, where people not only photograph or see UFOs but are hailed by the occupants and taken aboard. Some of these cases are useful to examine in retrospect. For example, Adamski was taken to the planet Venus, where conditions were very much like those in Eden. The extraterrestrials spoke mellifluously, walked among rivulets and flowers, wore long white robes, and gave heartening religious homilies.

We know now, as we did not know then, that the surface temperature of Venus is nine hundred degrees Fahrenheit. The surface pressures are ninety times what they are in this room. The atmosphere contains hydrochloric acid, hydrofluoric acid, and sulfuric acid. So at the very least, the long white robes would have been in tatters. We can in retrospect see that there must have been something wrong with the story. Maybe he just got the planet wrong. But one is left with the distinct impression that Adamski's account was contrived out of whole cloth.

It is remarkable that in all these million cases there is not one example of physical evidence that sustains even the most casual scrutiny. No pieces of spacecraft chipped off with a penknife and put into an envelope and carried back for laboratory examination of exotic alloys. No photograph of the interior of the spacecraft or the extraterrestrials, or a page from the captain's logbook. Somehow, in all of these cases, there is not a single example of concrete physical evidence. And that again is suggestive, I maintain, that we are dealing with some combination of psychopathology and conscious fraud and the misapprehension of natural phenomena, but not what is alleged by those who see UFOs.

I'd like to give you a specific case, because I think it's an example of how people with the best intentions in the world can nevertheless be badly fooled. Sometime in the 1950s, a highway patrolman in New Mexico is driving along a rural road that he knows extremely well, having driven along that road many, many times. And, to his astonishment, he sees a large, saucer-shaped object just settling down on the ground, the sunlight glinting off it. He's astonished. He pulls off to the side of the road and examines it. He then drives some tens of meters away to an emergency telephone at the side of the road and gets patched in to some scientists he happens to know at Los Alamos National Laboratory. He tells them, "The most extraordinary thing has just happened to me. This is a once-in-a-lifetime opportunity. I have just seen a flying saucer land. It is within my sight now. I have not had anything to drink. I am fully awake. I am in full possession of my senses. And if you get out here right away with monitoring equipment, we have the find of the century."

This scene is so compelling that the scientists are able to commandeer a helicopter and fly to the site. They land on the highway, approach the policeman—and, sure enough, in front of them is just what he described. Saucer-shaped, metallic, large, gleaming in the Sun. So, carrying their equipment, they rush toward it, and as they approach, they notice a farmer who is doing his farming things, totally oblivious to this large saucer that has just landed in front of him. They start thinking, is it possible that the saucer is invisible to the farmer but visible to them? Maybe the farmer has been hypnotized. They approach. The farmer finally sees them, if not the flying saucer, and challenges them. Why are they trespassing on his land? They say, "Because of the saucer." "Saucer? What saucer?" He turns around and looks exactly at it and apparently does not see it. Well, it turns out, after some few minutes of confused discussion, that what they were seeing was a silo for the storage of grain that the farmer was using, that he had himself made from—I've forgotten now from what, but it was indeed saucer-shaped—that he had been using for years.

Everything the highway patrolman had seen was right, except for one small detail. He had the impression that he had just seen it land, and he had not. Everything else was exactly as told. And what this stresses is that in an argument of this sort every link in the chain of argument has to be right. It's not enough for most links in the chain to be right. If you have one weak link, the entire chain of argument can collapse.

Now, it is sometimes said that people who take a skeptical approach to UFOs or ancient astronauts or indeed some varieties of revealed religion are engaging in prejudice. I maintain this is not prejudice. It is postjudice. That is, not a judgment made before examining the evidence but a judgment made after examining the evidence.

It does not say that as you finish reading this you will not walk outside and come upon a metallic flying saucer sitting there, posing embarrassment to the author. I would gladly trade my embarrassment for a genuine contact with an extraterrestrial civilization. But I maintain that after we have a certain amount of experience with such cases, an overall trend becomes clear, and that is that in cases of this sort we are enormously vulnerable to misunderstanding, to misevaluating. What we are talking about is not significantly different from what is called a miracle.

The definitive work on miracles was written by a famous Scottish philosopher, David Hume. In his book *An Inquiry Concerning Human Understanding,* in a famous chapter called "Of Miracles," Hume is considering a slightly but not very significantly different case.

> When anyone tells me that he saw a dead man restored to life, I immediately consider with myself whether it be more probable that this person should either deceive or be deceived or that the fact which he relates should really have happened. I weigh the one miracle against the other and according to the superiority which I discover, I pronounce my decision. Always I reject the greater miracle. If the falsehood of his testimony would be more miraculous than the event which he relates, then and not till then, can he pretend to command my belief or opinion.

And another way in which this has been phrased is by Thomas Paine, one of the heroes of the American Revolution, who is essentially paraphrasing Hume. He says, "Is it more probable that nature should go out of her course or that a man should tell a lie?"

What is being said here is that mere eyewitness testimony is insufficient if what is being reported is sufficiently extraordinary. Paine goes on to say,

> We have never seen, in our time, nature go out of her course. But we have good reason to believe that millions of lies have

been told in the same time. It is therefore at least millions to one that the reporter of a miracle tells a lie.

Strong stuff.

Without a doubt it is more interesting if miracles occur than if they do not. It makes a better story. And I can recall a case that happened to me. I was at a restaurant nearby Harvard University. Suddenly the proprietor and most of the diners rushed outside, napkins still tucked under their belts. My attention was attracted. I rushed outside also and saw a very strange light in the sky. I lived not far away, walked home (without paying the bill, but I told the proprietor I would return), got a pair of binoculars, came back, and with the binoculars was able to see that the one light was actually divided into two lights, that exterior to the two lights were a red light and a green light. The red light and the green light were blinking, and it was, it later turned out, a massive weather airplane with two powerful searchlights to determine the turbidity of the atmosphere. I told the people at the restaurant what I had seen. Everyone was uniformly disappointed. I asked why. And everyone had the same answer. It is a memorable story to go home and say, "I just saw a spaceship from another planet hovering over Harvard Square." It is a highly nonmemorable story to go home and say, "I saw an airplane with a bright light."

But beyond that, miracles speak to us of all sorts of things religious that we have powerful wishes to believe. This is true to such an extent that people become very angry when miracles are debunked. One of the most interesting cases of this sort—and there are thousands of them—is within the Roman Catholic Church, where there is an established procedure for verifying alleged miracles. It's in fact where the phrase "devil's advocate" comes from. The devil's advocate is the person who proposes alternative explanations of the alleged miracle, to see how good the evidence is. I have in front of me a newspaper clipping from June a year ago, titled "Priests Denounced After Rejecting Miracle Claim." And let me just read a few sentences:

Stockton, California. Angry believers denounced a panel of priests as "a bunch of devils" after the clergymen ruled that a weeping Madonna in a rural Roman Catholic church is probably a hoax, not a miracle. One woman, Lavergne Pita, burst into tears when the findings were announced Wednesday by the Diocese of Stockton. Manuel Pita protested that "these investigators are not investigators. They're a bunch of devils. How can they do this?" Reports that the sixty-pound statue sheds real tears and can move as far as thirty feet from its niche in Mater Ecclesiae Mission Church in Thornton began circulating two years ago. Church attendance has tripled since then. . . . Last year the diocese named a commission to study the reports. In announcing the panel's finding, Bishop Roger M. Mahoney said the events connected with the statue "do not meet the criteria for an authenticated appearance of Mary, the mother of Jesus Christ." The statue may have been moved, the tears may have been applied. . . . Actually, the tears were never reported to flow, they were just seen, and they were gluey. One of the proponents said, "When the virgin appeared to the kids in Portugal, they didn't believe them either. These things usually happen to the humble and low incomes. The poor," he added. "These things are not for everyone."

Well, I would like now to tell you about one of the most extraordinary studies on this subject that I know of, which is one of the few cases where not just supposed miraculous events occurred but where they were studied in great detail by a team of observers, who infiltrated the religious group in order to do sociological research. They convinced the group that they were there because they were also believers. This is an extremely interesting case, because the prophecies, every one of them, failed utterly. And those are not the cases we tend to hear.

The story comes from a book called *When Prophecy Fails,* by [Leon] Festinger et al. It was published in the middle 1960s and refers to events that occurred in Minneapolis, Minnesota, in the early 1950s. A woman in Minneapolis believed that she was receiving a message by automatic writing. Do you know what automatic writing is? It happens to people all over the world. It's where the hand with the pen or pencil in it seemingly takes on a life of its own and writes things when, as far as anyone else can see, the person who belongs to the hand is asleep or doing something else. There seems little doubt that the person who is attached to the hand is responsible for what is happening on the paper. But it has an eerie sense of happening not just unconsciously but from some external source. In this case the automatic writing was from Jesus—or at least a modern incarnation of him—who was resident on an otherwise undiscovered planet called Clarion. The message was urgent. It said that a flood would inundate the Earth (despite the biblical promise made to Noah), on the twenty-first of December, would cover most of the United States and the Soviet Union, among other nations, and would raise the lost continents of Atlantis and Mu. Spacemen from the planet Clarion would arrive before the flood to rescue the faithful, take them up on the flying saucers, and bring them to Clarion.

The group that formed around the woman who did the automatic writing were ordinary people, in no sense obviously deranged. One of the leaders of the group was a physician who was examined by psychiatrists, I guess on the grounds that for a physician to believe this was extraordinary but for anyone else it was expected. He was adjudged to be entirely sane although "holding unusual ideas." The group received numerous messages—six or eight—advising them to be present at a certain time in a certain place to be picked up by flying saucers before the event, and, as will be no surprise to you, the Clarionites never appeared. If they had appeared, you would have heard of it before now. The flood itself also never appeared, although earthquakes in several parts of the world occurred within a day of the predicted inundation, and that was taken by the enthusiasts in the group to be a partial confirmation of the flood.

As you can imagine, the failure of the flood on December 21 produced some consternation in the group but by no means led to the group falling apart. They responded wholeheartedly to a subsequent automatic-writing message that they were to sing Christmas carols in the cold outside the house of one of their leaders, preparatory to still another UFO pickup, which they did, surrounded by a crowd of some two hundred taunting onlookers and police to separate them from the onlookers. They showed great dedication, great courage. But a skeptical approach to the world, they cannot be said to have exhibited.

Now, as to their understanding of how it is that they were not picked up, there were several sets of explanations, and I'll just list them: They had misunderstood the message (although it said in plain English what they were to do and it was signed "Jesus" or "God Almighty"). Another explanation was that they had been insufficiently dedicated, that their faith had not been strong enough. Or that all this was merely a test by the extraterrestrials to see how committed they were and that the extraterrestrials never intended to flood the Earth, just to test their faith. Or that the predictions were entirely valid but they got the date wrong. It would happen ten thousand years later . . . a small mistake. Or that the inundation would have happened but the coterie of the faithful sufficiently impressed God with their faith that God intervened on behalf of mankind, and we're all alive because these people had believed strongly enough.

All these explanations are not mutually consistent, but they show a remarkable inventiveness and a striking unwillingness to change a set of beliefs in the face of contradictory evidence. Eventually most of the adherents drifted away from the movement, but even those who left first had repeatedly shown heroic fidelity in the face of what they call "disconfirmation," to say nothing of external skepticism. It's clear that mutual support within the belief system was central to the success, however short-lived, of the faith.

There was no charismatic leader here. No ambitious scoundrel. It was automatic writing and ordinary people. Indeed, the group cast about looking for guidance. They thought that spacemen from Clarion must be around them in the most unlikely contexts. For example, there were a bunch of leather-jacketed, motorcycle-riding young men who had come to scoff, whom they immediately took to be the angels from Clarion. And likewise the members of the social-science research team, who had infiltrated the movement trying to understand how religious movements get started, were also taken as angels from Clarion. This posed all sorts of challenges to the proper detachment of scientist from subject.

Most of these people had previously been involved in other borderline religions or pseudoscientific groups, including UFO clubs, spiritualists, Dianetics, which has since transmogrified into something called Scientology, and so on. But it is the very ordinariness of this group that I believe gives some real insights into the origins of religion. Let me quote the concluding sentences by Festinger et al.:

They were unskilled proselytizers. It is interesting to speculate, however, on what they might have made of their opportunities had they been more effective apostles. For about a week they were headline news throughout the nation. Their ideas were not without popular appeal and they received hundreds of visitors, telephone calls and letters from seriously interested citizens as well as offers of money which they invariably refused. Events conspired to offer them a truly magnificent opportunity to grow in numbers. Had they been more effective, disconfirmation might have portended the beginning and not the end.

Suppose they'd had a charismatic leader. Or suppose that by chance there had been a spectacular UFO sighting at the time of the predicted inundation, for example, an Air Force test of a new kind of aircraft. Or suppose that the message that came from Clarion was not just that there was going to be a flood but something powerful, something moving, something that spoke to an oppressed minority in the United States or elsewhere. Then I think we can see the possibility that the Clarion religion would have grown into something much larger. If we look at recent religions—and let me restrict myself to those that have more than a million adherents—we find, for example, one that confidently predicted that the world would end in 1914. Unambiguous. And when the world did not end in 1914 (as far as one can tell it has not), they did not argue that, oh, they made a small mistake in arithmetic, it was actually 2014, hope no one was inconvenienced. They did not say that, well, the world *would* have ended, but they were sufficiently faithful that God intervened. No. They said, and it is still the major tenet of their faith, that the world *did* end in 1914 and we simply haven't noticed yet. This is a religion with millions of adherents, currently in the United States.

Or there is a religion that says that all diseases are psychogenic, that there is no such thing as a microorganism producing disease. There is no such thing as a cellular malfunction producing a disease, that the only thing that produces disease is not thinking right, not having adequate faith. And I need not remind you that there is a significant body of medical evidence to the contrary.

There is a religion that believes that in the nineteenth century a set of golden tablets was prepared by an angel and dug up by a divinely inspired human being. And the tablets were written in ancient Egyptian hieroglyphics and had on them a hitherto-unknown set of books like those in the Old Testament. And, unfortunately, the tablets are not available for any scrutiny these days, and in addition there is powerful evidence of conscious fraud at the time that the religion was founded, which led, last week, to two people being killed in the state of Utah, having to do with some early letters from the founders of the religion that were inconsistent with doctrine.

Or there is a religion that believes that if you only have enough faith, you can levitate. I mean, that you can bodily float off the ground and propel yourself. It has many practical applications, if only it were true. These are perfectly typical tenets or aspects of modern religions.

And if that is true, what about ancient religions? After all, there is a much greater distance in time between us and those earlier religions. And that means that there are much larger opportunities for fraud and for changing the disquieting details. I remind you that rewriting history is done all the time. To give an example—there are so many—one of the leaders of the

Russian Revolution was a man named Lev Davidovich Bronstein, also known as Leon Trotsky. He founded the Red Army, he established the modern Soviet railroad system, he was the founder and first editor of *Pravda*, he played a leading role in both the 1905 and the 1917 revolutions, but he does not exist in the Soviet Union. He's not there. You cannot find anything about him. There is no picture of him. In a two-volume Soviet history of the world, he appears once, as having inappropriate agricultural views. Otherwise unmentioned. They have simply written him out of the history of their own revolution, in which he played an absolutely central role, second perhaps only to that of Lenin. So now imagine that a religion is founded not just a few decades ago but a few centuries or a few thousand years ago, in which the received wisdom passes through a small group—a small priesthood. Think of the opportunities for changing disquieting facts in the interim. David Hume says,

> The many instances of forged miracles and prophecies and supernatural events, which in all ages have either been detected by contrary evidence or which detect themselves by their absurdity, prove sufficiently the strong propensity of mankind to the extraordinary and marvelous and ought reasonably to beget a suspicion against all relations of this kind. It is strange, a judicious reader is apt to say, that such prodigious events never happen in our day, but it is nothing strange that men should lie in all ages.

And then on the point that I was just making, he says,

> In the infancy of new religions the wise and learned commonly esteem the matter too inconsiderable to deserve their attention or regard. And then when afterwards they would

willingly detect the cheat in order to undeceive the deluded multitudes, the season is now past and the records and witnesses which might clear up the matter have perished beyond recovery.

Well, it seems to me that there is only one conceivable approach to these matters. If we have such an emotional stake in the answers, if we want badly to believe, and if it is important to know the truth, then nothing other than a committed, skeptical scrutiny is required. It is not very different from buying a used car. When you buy a used car, it is insufficient to remember that you badly need a car. After all, it has to work. It is insufficient to say that the used-car salesman is a friendly fellow. What you generally do is you kick the tires, you look at the odometer, you open up the hood. If you do not feel yourself expert in automobile engines, you bring a friend who is. And you do this for something as unimportant as an automobile. But on issues of the transcendent, of ethics and morals, of the origin of the world, of the nature of human beings, on those issues should we not insist upon at least equally skeptical scrutiny?

S i x

THE GOD HYPOTHESIS

The Gifford Lectures are supposed to be on the topic of natural theology. Natural theology has long been understood to mean theological knowledge that can be established by reason and experience and experiment alone. Not by revelation, not by mystical experience, but by reason. And this is, in the long, historical sweep of the human species, a reasonably novel view. For example, we might look at the following sentence written by Leonardo da Vinci. In his notebooks he says, "Whoever in discussion adduces authority uses not intellect but rather memory."

This was an extremely heterodox remark for the early sixteenth century, when most knowledge was derived from authority. Leonardo himself had many clashes of this sort. During a trip to an Apennine mountaintop, he had discovered the fossilized remains of shellfish that ordinarily lived on the ocean floor. How did this come about? The conventional theological wisdom was that the Great Flood of Noah had inundated the mountaintops and carried the clams and oysters with it. Leonardo, remembering that the Bible says that the flood lasted only forty days, attempted to calculate whether this would be sufficient time to carry the shellfish with them, even if the mountaintops were inundated. During what state in the life cycle of the shellfish had they been deposited?—and so on. He came to the conclusion this didn't work, and proposed a quite daring alternative; namely, that over immense vistas of geological time the mountaintops had pushed up through the oceans. And that posed all sorts of theological difficulties. But it is the correct answer, as I think it's fair to say it has been definitively established in our time.

If we are to discuss the idea of God and be restricted to rational arguments, then it is probably useful to know what we are talking about when we say "God." This turns out not to be easy. The Romans called the Christians atheists. Why? Well, the Christians had a god of sorts, but it wasn't a real god. They didn't believe in the divinity of apotheosized emperors or Olympian gods. They had a peculiar, different kind of god. So it was very easy to call people who believed in a different kind of god atheists. And that general sense that an atheist is anybody who doesn't believe exactly as I do prevails in our own time.

Now, there is a constellation of properties that we generally think of when we in the West, or more generally in the Judeo-Christian-Islamic tradition, think of God. The fundamental differences among Judaism, Christianity, and Islam are trivial compared to their similarities. We think of some being who is omnipotent, omniscient, compassionate, who created the universe, is responsive to prayer, intervenes in human affairs, and so on.

But suppose there were definitive proof of some being who had some but not all of these properties. Suppose somehow it were demonstrated that there was a being who originated the universe but is indifferent to prayer. . . . Or, worse, a god who was oblivious to the existence of humans. That's very much like Aristotle's god. Would that be God or not? Suppose it were someone who was omnipotent but not omniscient, or vice versa. Suppose this god understood all the consequences of his actions but there were many things he was unable to do, so he was condemned to a universe in which his desired ends could not be accomplished. These alternative kinds of gods are hardly ever thought about or discussed. A priori there is no reason they should not be as likely as the more conventional sorts of gods.

And the subject is further confused by the fact that prominent theologians such as Paul Tillich, for example, who gave the Gifford Lectures many years ago, explicitly denied God's existence, at least as a supernatural power. Well, if an esteemed theologian (and he's by no means the only one) denies that God is a supernatural being, the subject seems to me to be somewhat confused. The range of hypotheses that are seriously covered under the rubric "God" is immense. A naive Western view of God is an outsize, light-skinned male with a long white beard, who sits on a very large throne in the sky and tallies the fall of every sparrow.

Contrast this with a quite different vision of God, one proposed by Baruch Spinoza and by Albert Einstein. And this second kind of god they called God in a very straightforward way. Einstein was constantly interpreting the world in terms of what God would or wouldn't do. But by God they meant something not very different from the sum total of the physical laws of the universe; that is, gravitation plus quantum mechanics plus grand unified field theories plus a few other things equaled God. And by that all they meant was that here were a set of exquisitely powerful physical principles that seemed to explain a great deal that was otherwise inexplicable about the universe. Laws of nature, as I have said earlier, that apply not just locally, not just in Glasgow, but far beyond: Edinburgh, Moscow, Peking, Mars, Alpha Centauri, the center of the Milky Way, and out by the most distant quasars known. That the same laws of physics apply everywhere is quite remarkable. Certainly that represents a power greater than any of us. It represents an unexpected regularity to the universe. It need not have been. It could have been that every province of the cosmos had its own laws of nature. It's not apparent from the start that the same laws have to apply everywhere.

Now, it would be wholly foolish to deny the existence of laws of nature. And if that is what we are talking about when we say God, then no one can possibly be an atheist, or at least anyone who would profess atheism would have to give a coherent argument about why the laws of nature are inapplicable.

I think he or she would be hard-pressed. So with this latter definition of God, we all believe in God. The former definition of God is much more dubious. And there is a wide range of other sorts of gods. And in every case we have to ask, "What kind of god are you talking about, and what is the evidence that this god exists?"

Certainly if we are restricted to natural theology, it is insufficient to say, "I believe in that sort of god, because that's what I was told when I was young," because other people are told different things about quite different religions that contradict those of my parents. So they can't all be right. And in fact they all may be wrong. It is certainly true that many different religions are mutually inconsistent. It's not that they just aren't perfect simulacrums of each other but rather that they grossly contradict each other.

I'll give you a simple example; there are many. In the Judeo-Christian-Islamic tradition, the world is a finite number of years old. By counting up the begats in the Old Testament, you can come to the conclusion that the world is a good deal less than ten thousand years old. In the seventeenth century, the archbishop of Armagh, James Ussher, made a courageous but fundamentally flawed effort to count them up precisely. He came to a specific date on which God created the world. It was October 23 in 4004 B.C., a Sunday.

Now, think again of all the possibilities: worlds without gods; gods without worlds; gods that are made by preexisting gods; gods that were always here; gods that never die; gods that do die; gods that die more than once; different degrees of divine intervention in human affairs; zero, one, or many prophets; zero, one, or many saviors; zero, one, or many resurrections; zero, one, or many gods. And related questions about sacrament, religious mutilation, and scarification, baptism, monastic orders, ascetic expectations, the presence or absence of an afterlife, days to eat fish, days not to eat at all, how many afterlives you have coming to you, justice in this world or the next world or no world at all, reincarnation, human sacrifice, temple prostitution, jihads, and so forth. It's a vast array of things that people believe. Different religions believe different things. There's a grab bag of religious alternatives. And there are clearly more combinations of alternatives than there are religions, even though there are something like a few thousand religions on the planet today. In the history of the world, there probably were many tens, maybe hundreds of thousands, if you think back to our hunter-gatherer ancestors when the typical human community was a hundred or so people. Back then there were as many religions as there were hunter-gatherer bands, although the differences between them were probably not all that great. But nobody knows, since, unfortunately, we have virtually no knowledge left of what our ancestors for the greatest part of the tenure of humans on this planet believed, because word-of-mouth tradition is inadequate and writing had not been invented.

So, considering this range of alternatives, one thing that

comes to my mind is how striking it is that when someone has a religious-conversion experience, it is almost always to the religion or one of the religions that are mainly believed in his or her community. Because there are so many other possibilities. For example, it's very rare in the West that someone has a religious-conversion experience in which the principal deity has the head of an elephant and is painted blue. That is quite rare. But in India there is a blue, elephant-headed god that has many devotees. And seeing depictions of this god there is not so rare. How is it that the apparition of elephant gods is restricted to Indians and doesn't happen except in places where there is a strong Indian tradition? How is that apparitions of the Virgin Mary are common in the West but rarely occur in places in the East where there isn't a strong Christian tradition? Why don't the details of the religious belief cross over the cultural barriers? It is hard to explain unless the details are entirely determined by the local culture and have nothing to do with something that is externally valid.

Put another way, any preexisting predisposition to religious belief can be powerfully influenced by the indigenous culture, wherever you happen to grow up. And especially if the children are exposed early to a particular set of doctrine and music and art and ritual, then it is as natural as breathing, which is why religions make such a large effort to attract the very young.

Or let's take another possibility. Suppose a new prophet arises who claims a revelation from God, and that revelation contravenes the revelations of all previous religions. How is the average person, someone not so fortunate as to have received this revelation personally, to decide whether this new revelation is valid or not? The only dependable way is through natural theology. You have to ask, "What is the evidence?" And it's insufficient to say, "Well, there is this extremely charismatic person who said that he had a conversion experience." Not enough. There are lots of charismatic people who have all sorts of mutually exclusive conversion experiences. They can't all be right. Some of them have to be wrong. Many of them have to be wrong. It's even possible that all of them are wrong. We cannot depend entirely on what people say. We have to look at what the evidence is.

I would like now to turn to the issue of alleged evidence or, as they're called, proofs of the existence of God. And I will mainly spend my time on the Western proofs. But to show an ecumenical spirit, let me begin with some Hindu proofs, which in many ways are as sophisticated and certainly more ancient than the Western arguments.

Udayana, an eleventh-century logician, had a set of seven proofs of the existence of God, and I won't mention all of them; I'll just try to convey a sense of it. And, by the way, the kind of god that Udayana is talking about is not exactly the same, as you might imagine, as the Judeo-Christian-Islamic god. His god is all-knowing and imperishable but not necessarily omnipotent and compassionate.

First, Udayana reasons that all things must have a cause. The world is full of things. Something must have made those things. And this is very similar to a Western argument that we'll come to shortly.

Secondly, an argument not heard in the West is the argument from atomic combinations. It is quite sophisticated. It says at the beginning of Creation, atoms had to be bonded with each other to make bigger things. And such a bonding of atoms always requires the activity of a conscious agent. Well, now we know that's false. Or we know, at least, that there are laws of atomic interaction that determine how atoms bind together. It's a subject called chemistry. And you might say that this is due to the intervention of a deity but it does not require the direct intervention of a deity. All the deity has to do is establish the laws of chemistry and retire.

Third is an argument from the suspension of the world. The world isn't falling, as is clear by just looking out. We're not hurtling through space, apparently, and therefore something is holding the world up, and that something is God. Well, this is a quite natural view of things. It's connected with the idea that we are stationary and at the center of the universe, a misapprehension that all peoples all over the world have had. In fact we are falling at a terrific rate of speed in orbit around the Sun. And every year we go 2 pi times the radius of the Earth orbit. If you work that out, you'll find it's extremely fast.

Fourth is an argument from the existence of human skills. And this is very close to the von Däniken argument that if someone didn't show us how to do things, we wouldn't know how to do it. I think there's plentiful argument against that.

Then there is the existence of authoritative knowledge separate from human skills. How would we know things that are in, for example, the Vedas, the Hindu holy books, unless God had written them? The idea that humans were able to write the Vedas was difficult for Udayana to accept.

Well, this gives a sense of these arguments and shows that there is a pervasive human wish to give a rational explanation for the existence of a God or gods, and also, I maintain, it demonstrates that these arguments are not always highly successful. Let me now go to some of the Western arguments, which may be entirely familiar to everyone, in which case I apologize.

First of all, there is the cosmological argument, which is not very different from the argument we just heard. The cosmological argument in the West essentially has to do with causality.

There are things all around us; those things were caused by something else. And so, after a while, you find yourself back to remote times and causes. Well, it can't go on forever, an infinite regress of causes, as Aristotle and later Thomas Aquinas argued, and therefore you need to come to an uncaused first cause. Something that started everything going that was not itself caused; that is, that was always there. And this is defined as God.

There are two conflicting hypotheses here, two alternative hypotheses. One is that the universe was always here, and the other is that God was always here. Why is it immediately obvious that one of these is more likely than the other? Or, put another way, if we say that God made the universe, it is reasonable to then ask, "And who made God?"

Virtually every child asks that question and is usually shushed by the parents and told not to ask embarrassing questions. But how does saying that God made the universe, and never mind asking where God came from, how is that more satisfying than to say the universe was always here?

In modern astrophysics there are two contending views. First of all, there is no doubt in my mind, and I think almost all astrophysicists agree, that the evidence from the expansion of the universe, the mutual recession of the galaxies and from what is called the three-degree black-body background radiation, suggests that something like 13 or 15 billion years ago all the matter in the universe was compressed into an extremely small volume, and that something that can surely be called an explosion happened at that time, and that the subsequent expansion of the universe and the condensation of matter led to galaxies, stars, planets, living beings, and all the rest of the details of the universe we see around us.

Now, what happened before that? There are two views. One

is "Don't ask that question," which is very close to saying that God did it. And the other is that we live in an oscillating universe in which there is an infinite number of expansions and contractions.*

We happen to be roughly 15,000 million years out from the last expansion. And some, let's say, 80,000 million years from now, the expansion will stop, to be replaced by a compression, and all the matter will fly together to a very small volume and then expand again with no information trickling through the cusps in the expansion process.

The former of these views happens, by chance, to be close to the Judeo-Christian-Islamic view, the latter close to the standard Hindu views. And so, if you like, you can think of the varying contentions of these two major religious views being fought out on the field of contemporary satellite astronomy. Because that's where the answer to this question will very likely be decided. Is there enough matter in the universe to prevent the expansion from continuing forever, so that the self-gravity will make the expansion stop and be followed by a contraction? Or is there not enough matter in the universe to stop the expansion, so everything keeps expanding forever? This is an experimental question. And it is very likely that in our lifetime we will have the answer to it. And I stress that this is very different from the usual theological approach, where there is never an experiment that can be performed to test out any contentious issue. Here there is one. So we don't have to make judgments now. All we have to do is maintain some tolerance for ambiguity until the data are in, which may happen in a decade or less. It is possible

---

*In 1998 two international teams of astronomers independently reported unexpected evidence that the expansion of the universe is accelerating. These findings suggest that the universe is not oscillating but will continue to expand forever.

that the Hubble Space Telescope, scheduled for launch next summer, will provide the answer to this question. It's not guaranteed but it is possible.*

Now, by the way, on this issue of who's older, God or the universe, there's actually a three-by-three matrix: God can have always existed but will not exist for all future time. That is to say God might have no beginning but might have an end. God might have a beginning but no end. God might have no beginning and no end. Likewise for the universe. The universe might be infinitely old, but it will end. The universe might have begun a finite time ago but will go on forever, or it might have always existed and will never end. Those are just the logical possibilities. And it's curious that human myth has some of those possibilities but not others. I think in the West it's quite clear that there is a human or animal life-cycle model that has been imposed on the cosmos. It's a natural thing to think about, but after a while its limitations, I think, become clear.

Also, I should say something about the Second Law of Thermodynamics. An argument that is sometimes used to justify a belief in God is that the Second Law of Thermodynamics says that the universe as a whole runs down; that is, the net amount of order in the universe must decline. Chaos must increase as time goes on; that is, in the entire universe. It doesn't say that in a given locale, such as the Earth, the amount of order can't increase, and clearly it has. Living things are much more complex, have much more order in them, than the raw materials from which life formed some 4,000 million years ago. But this increase in order on the Earth is done, it is easy enough to calculate, at the expense of a decrease in order on the Sun, which is the source of the energy that drives terrestrial biology. It's by no

* Earth-based telescopes provided the answer in 1998. See previous note.

means clear, by the way, that the Second Law of Thermodynamics applies to the universe as a whole, because it is an experimental law, and we don't have experience with the universe as a whole. But it's always struck me as curious that those who wish to apply the Second Law to theological issues do not ask whether God is subject to the Second Law. Because if God were subject to the Second Law of Thermodynamics, then God could have only a finite lifetime. And again, there is an asymmetric use of the principles of physics when theology confronts thermodynamics.

Also, by the way, if there were an uncaused first cause, that by no means says anything about omnipotence or omniscience, or compassion, or even monotheism. And Aristotle, in fact, deduced several dozen first causes in his theology.

The second standard Western argument using reason for God is the so-called argument from design, which we have already talked about, both in its biological context and in the recent astrophysical incarnation called the anthropic principle. It is at best an argument from analogy; that is, that some things were made by humans and now here is something more complex that wasn't made by us, so maybe it was made by an intelligent being smarter than us. Well, maybe, but that is not a compelling argument. I tried to stress earlier the extent to which misunderstandings, failure of the imagination, and especially the lack of awareness of new underlying principles may lead us into error with the argument from design. The extraordinary insights of Charles Darwin on the biological end of the argument of design provide clear warning that there may be principles that we do not yet divine (if I may use that word) underlying apparent order.

There is certainly a lot of order in the universe, but there is also a lot of chaos. The centers of galaxies routinely explode, and if there are inhabited worlds and civilizations there, they are destroyed by the millions, with each explosion of the galactic nucleus or a quasar. That does not sound very much like a god who knows what he, she, or it is doing. It sounds more like an apprentice god in over his head. Maybe they start them out at the centers of galaxies and then after a while, when they get some experience, move them on to more important assignments.

Then there is the moral argument for the existence of God generally attributed to Immanuel Kant, who was very good at showing the deficiencies of some of the other arguments. Kant's argument is very simple. It's just that we are moral beings; therefore God exists. That is, how else would we know to be moral?

Well, first of all you might argue that the premise is dubious. The degree to which humans can be said to be moral beings without the existence of some police force is open at least to debate. But let's put that aside for the moment. Many animals have codes of behavior. Altruism, incest taboos, compassion for the young, you find in all sorts of animals. Nile crocodiles carry their eggs in their mouths for enormous distances to protect the young. They could make an omelette out of it, but they choose not to do so. Why not? Because those crocodiles who enjoy eating the eggs of their young leave no offspring. And after a while all you have is crocodiles who know how to take care of the young. It's very easy to see. And yet we have a sense of thinking of that as being somehow ethical behavior. I'm not against taking care of children; I'm strongly for it. All I'm saying is, it does not follow if we are powerfully motivated to take care of our young or the young of everybody on the planet, that God made us do it. Natural selection can make us do it, and almost surely has. What's more, once humans reach the point of awareness of their surroundings, we can figure things out, and we can see what's good for our own survival as a community or a nation or a species and take steps to ensure our survival. It's not hopelessly beyond our ability. It's not clear to me that this requires the existence of God to explain the limited but definite degree of moral and ethical behavior that is apparent in human society.

Then there is the curious argument, unique to the West, called the "ontological argument," which is generally associated with [St.] Anselm, who died in 1109. His argument can be very simply stated: God is perfect. Existence is an essential attribute of perfection. Therefore, God exists. Got it? I'll say it again. God is perfect. Existence is an essential attribute of perfection. You can't be perfect if you don't exist, Anselm says. Therefore God exists. While this argument has for brief moments captured very significant thinkers (Bertrand Russell describes how it suddenly hit him that Anselm might be right—for about fifteen minutes), this is not considered a successful argument. The twentieth-century logician Ernest Nagel described it as "confounding grammar with logic."

What does it mean, "God is perfect"? You need a separate description of what constitutes perfection. It's not enough to say "perfect" and do not ask what "perfect" means. And how do you know God is perfect? Maybe that's not the god that exists, the perfect one. Maybe it's only imperfect ones that exist. And then why is it that existence is an essential attribute of perfection? Why isn't nonexistence an essential attribute of perfection? We are talking words. In fact, there is the remark that is sometimes made about Buddhism, I think in a kindly light, that their god is so great he doesn't even have to exist. And that is the perfect counterpoise to the ontological argument. In any case, I do not think that the ontological argument is compelling.

Then there's the argument from consciousness. I think,

therefore, God exists; that is, how could consciousness come into being? And, indeed, we do not know the details in any but the very broadest brush about the evolution of consciousness. That is on the agenda of future neurological science. But we do know, for example, that an earthworm introduced into a Y-shaped glass tube with, let's say, an electric shock on the right-hand fork and food in the left-hand fork, rapidly learns to take the left fork. Does an earthworm have consciousness if it is able after a certain number of trials invariably to know where the food is and the shock isn't? And if an earthworm has consciousness, could a protozoan have consciousness? Many phototropic microorganisms know to go to the light. They have some kind of internal perception of where the light is, and nobody taught them that it's good to go to the light. They had that information in their hereditary material. It's encoded into their genes and chromosomes. Well, did God put that information there, or might it have evolved through natural selection?

It is clearly good for the survival of microorganisms to know where the light is, especially the ones that photosynthesize. It is certainly good for earthworms to know where the food is. Those earthworms that can't figure out where the food is leave few offspring. After a while the ones that survive know where the food is. Those phototropic or phototactic offspring have encoded into their genetic material how to find the light. It is not apparent that God has entered into the process. Maybe, but it's not a compelling argument. And the general view of many, not all, neurobiologists is that consciousness is a function of the number and complexity of neuronal linkages of the architecture of the brain. Human consciousness is what happens when you get to something like 10^{11} neurons and 10^{14} synapses. This raises all sorts of other questions: What is consciousness like when you have 10^{20}

synapses or 10^{30}? What would such a being have to say to us any more than we would have to say to the ants? So at least it does not seem to me that the argument from consciousness, a continuum of consciousness running through the animal and plant kingdoms, proves the existence of God. We have an alternative explanation that seems to work pretty well. We don't know the details, although work on artificial intelligence may help to clarify that. But we don't know the details of the alternative hypothesis either. So it could hardly be said that this is compelling.

Then there's the argument from experience. People have religious experiences. No question about it. They have them worldwide, and there are some interesting similarities in the religious experiences that are had worldwide. They are powerful, emotionally extremely convincing, and they often lead to people reforming their lives and doing good works, although the opposite also happens. Now, what about this? Well, I do not mean in any way to object to or deride religious experiences. But the question is, can any such experience provide other than anecdotal evidence of the existence of God or gods? One million UFO cases since 1947. And yet, as far as we can tell, they do not correspond—any of them—to visitations to the Earth by spacecraft from elsewhere. Large numbers of people can have experiences that can be profound and moving and still not correspond to anything like an exact sense of external reality. And the same can be said not just about UFOs but about extrasensory perception and ghosts and leprechauns and so on. Every culture has things of this sort. That doesn't mean that they all exist; it doesn't mean that any of them exist.

I also note that religious experiences can be brought on by specific molecules. There are many cultures that consciously imbibe or ingest those molecules in order to bring on a religious experience. The peyote cult of some Native Americans is exactly that, as is the use of wine as a sacrament in many Western religions. It's a very long list of materials that are taken by humans in order to produce a religious experience. This suggests that there is some molecular basis for the religious experience and that it need not correspond to some external reality. I think it's a fairly central point—that religious experiences, personal religious experiences, not the natural theological evidence for God, if any, can be brought on by molecules of finite complexity.

So if I then run through these arguments—the cosmological argument, the argument from design, the moral argument, the ontological argument, the argument from consciousness, and the argument from experience—I must say that the net result is not very impressive. It is very much as if we are seeking a rational justification for something that we otherwise hope will be true.

And then there are certain classical problems with the existence of God. Let me mention a few of them. One is the famous problem of evil. This basically goes as follows: Grant for a moment that evil exists in the world and that unjust actions sometimes go unpunished. And grant also that there is a God that is benevolent toward human beings, omniscient, and omnipotent. This God loves justice, this God observes all human actions, and this God is capable of intervening decisively in human affairs. Well, it was understood by the pre-Socratic philosophers that all four of these propositions cannot simultaneously be true. At least one has to be false. Let me say again what they are. That evil exists, that God is benevolent, that God is omniscient, that God is omnipotent. Let's just see about each of them.

First of all, you might say, "Well, evil doesn't exist in the world. We can't see the big picture, that a little pool of evil here is awash in a great sea of good that it makes possible." Or, as medieval theologians used to say, "God uses the Devil for his own purposes." This is clearly the three-monkey argument about "hear no evil . . ." and has been described by a leading contemporary theologian as a gratuitous insult to mankind, a symptom of insensitivity and indifference to human suffering. To be assured that all the miseries and agonies men and women experience are only illusory. Pretty strong.

This is clearly hoping that the disquieting facts go away if you merely call them something else. It is argued that some pain is necessary for a greater good. But why, exactly? If God is omnipotent, why can't He arrange it so there is no pain? It seems to me a very telling point.

The other alternatives are that God is not benevolent or compassionate. Epicurus held that God was okay but that humans were the least of His worries. There are a number of Eastern religions that have something like that same flavor. Or God isn't omniscient; He doesn't know everything; He has business elsewhere and so doesn't know that humans are in trouble. One way to think about it is there are several times 10^{11} worlds in every galaxy and several times 10^{11} galaxies, and God's busy.

The other possibility is that God isn't omnipotent. He can't do everything. He could maybe start the Earth off or create life, intervene occasionally in human history, but can't be bothered day in and day out to set things right here on Earth. Now, I don't claim to know which of these four possibilities is right, but it's clear that there is a fundamental contradiction at the heart of the Western theological view produced by the problem of evil. And I've read an account of a recent theological conference devoted to this problem, and it clearly was an embarrassment to the assembled theologians.

This raises an additional question—a related question—and that has to do with microintervention. Why in any case is it necessary for God to intervene in human history, in human affairs, as almost every religion assumes happens? That God or the gods come down and tell humans, "No, don't do that, do this, don't forget this, don't pray in this way, don't worship anybody else, mutilate your children as follows." Why is there such a long list of things that God tells people to do? Why didn't God do it right in the first place? You start out the universe, you can do anything. You can see all future consequences of your present action. You want a certain desired end. Why don't you arrange it in the beginning? The intervention of God in human affairs speaks of incompetence. I don't say incompetence on a human scale. Clearly all of the views of God are much more competent than the most competent human. But it does not speak of omnicompetence. It says there are limitations.

I therefore conclude that the alleged natural theological arguments for the existence of God, the sort we're talking about, simply are not very compelling. They are trotting after the emotions, hoping to keep up. But they do not provide any satisfactory argument on their own. And yet it is perfectly possible to imagine that God, not an omnipotent or an omniscient god, just a reasonably competent god, could have made absolutely clearcut evidence of His existence. Let me give a few examples.

Imagine that there is a set of holy books in all cultures in which there are a few enigmatic phrases that God or the gods tell our ancestors are to be passed on to the future with no change. Very important to get it exactly right. Now, so far that's not very different from the actual circumstances of alleged holy books. But suppose that the phrases in question were phrases that we would recognize today that could not have been recognized then. Simple example: The Sun is a star. Now, nobody knew that, let's say, in the sixth century B.C., when the Jews were in the Babylonian exile and picked up the Babylonian cosmology from the principal astronomers of the time. Ancient Babylonian science is the cosmology that is still enshrined in the book of Genesis. Suppose instead the story was "Don't forget, the Sun is a star." Or "Don't forget, Mars is a rusty place with volcanoes. Mars, you know, that red star? That's a world. It has volcanoes, it's rusty, there are clouds, there used to be rivers. There aren't anymore. You'll understand this later. Trust me. Right now, don't forget."

Or, "A body in motion tends to remain in motion. Don't think that bodies have to be moved to keep going. It's just the opposite, really. So later on you'll understand that if you didn't have friction, a moving object would just keep moving." Now, we can imagine the patriarchs scratching their heads in bewilderment, but after all it's God telling them. So they would copy it down dutifully, and this would be one of the many mysteries in holy books that would then go on to the future until we could recognize the truth, realize that no one back then could possibly have figured it out, and therefore deduce the existence of God.

There are many cases that you can imagine like this. How about "Thou shalt not travel faster than light"? Okay, you might argue that nobody was at imminent risk of breaking that commandment. It would have been a curiosity: "We don't understand what that one's about, but all the others we abide by." Or "There are no privileged frames of reference." Or how about some equations? Maxwell's laws in Egyptian hieroglyphics or ancient Chinese characters or ancient Hebrew. And all the terms are defined: "This is the electric field, this is the magnetic field." We don't know what those are, but we'll just copy them down, and then later, sure enough, it's Maxwell's laws or the Schrödinger equation. Anything like that would have been possible had God existed and had God wanted us to have evidence of His existence. Or in biology. How about, "Two strands entwined is the secret of life"? You may say that the Greeks were onto that because of the caduceus. You know, in the American army all the physicians wore the caduceus on their lapels, and various medical insurance schemes also use it. And it is connected with, if not the existence of life, at least saving it. But there are very few people who use this to say that the correct religion is the religion of the ancient Greeks, because they had the one symbol that survives critical scrutiny later on.

This business of proofs of God, had God wished to give us some, need not be restricted to this somewhat questionable method of making enigmatic statements to ancient sages and hoping they would survive. God could have engraved the Ten Commandments on the Moon. Large. Ten kilometers across per commandment. And nobody could see it from the Earth but then one day large telescopes would be invented or spacecraft would approach the Moon, and there it would be, engraved on the lunar surface. People would say, "How could that have gotten there?" And then there would be various hypotheses, most of which would be extremely interesting.

Or why not a hundred-kilometer crucifix in Earth orbit? God could certainly do that. Right? Certainly, create the universe? A simple thing like putting a crucifix in Earth orbit? Perfectly possible. Why didn't God do things of that sort? Or, put another way, why should God be so clear in the Bible and so obscure in the world?

I think this is a serious issue. If we believe, as most of the great theologians hold, that religious truth occurs only when there is a convergence between our knowledge of the natural world and revelation, why is it that this convergence is so feeble when it could easily have been so robust?

So, to conclude, I would like to quote from Protagoras in the fifth century B.C., the opening lines of his *Essay on the Gods:*

> About the gods I have no means of knowing either that they exist or that they do not exist or what they are to look at. Many things prevent my knowing. Among others, the fact that they are never seen.

THE RELIGIOUS EXPERIENCE

Cast your mind back some hundreds of thousands of years. Those who can do that readily will have demonstrated some of the issues that I considered dubious earlier, but apart from reincarnation let's try to think about what were the circumstances of the greater part of the tenure of the human species on Earth. That surely is relevant to any attempt to understand our present circumstances.

The human family is some millions of years old, the human species perhaps one million, with some uncertainty. For the greater part of that period by far, we did not have anything like present technology, present social organization, or present religions. And yet our emotional predispositions were powerfully set in those times. Whatever our feelings and thoughts and approaches to the world were then, they must have been selectively advantageous, because we have done rather well. On this planet we are certainly the dominant organism of some fair size. An argument could be made for beetles or bacteria at smaller scales as being the dominant organism on the planet, but at least on our scale we have done quite well.

Now, what were those characteristics, and how would we know what they are? Well, one way we can know is by examining the groups of hunter-gatherers that are still tenuously alive on the planet today. These are small groups of people whose way of life predates the invention of agriculture. The fact that we know them means they must have made some contact with our present global civilization—and that immediately implies that their way of life is in its last days. They are the essence of humans. They have been studied by dedicated anthropologists who have lived with them, learned their languages, been adopted into the group in those cases that permit outsiders to have such an experience, and we can learn something about them. They are by no means all the same. This is a large topic, called cultural anthropology. I do not pretend to be expert in it, but I have had the benefit of spending a fair amount of time with some of the anthropologists who have been at the forefront of studying some of these groups. And I think it's relevant to the task before us.

There are, as I say, different kinds of groups, including some that we might consider absolutely horrendous and some that we might consider astonishingly benign, and I'll try to give a sense of each.

For the latter let me say just a few words about the !Kung people in the Kalahari Desert in the Republic of Botswana. These are a people who now have been drafted into the army of apartheid South Africa, and their culture has been irrevocably abused. But up until some twenty years ago, they had been well studied. We know something about them.

They are hunter-gatherers, which mainly means that the men hunt and the women gather. There is a kind of sexual division of labor, but there is very little in the way of social hierarchy. There is not a significant male dominance of women. In fact, there's very little in the way of social hierarchy at all. There is specialization of labor. That's different from social hierarchy. Children are treated with tenderness and understanding. And there is very little in the way of warfare, although occasionally they run into difficulties because of misunderstandings.

For example, there was a famous case, sometime ago, in which a hunting party came back and said that there was the most astonishing good fortune—a completely new creature had been discovered, and you could actually creep up to it with your bow and arrow and get within a meter of it, and it would not run away. And then you could shoot it dead. And here it is. And it was a cow. The neighboring Herero people protested, and this conflict between two groups, one that had not yet left the hunter-gatherer stage and the other that had domesticated animals, then had to be settled.

Another interesting question has to do with the hunt. Who owns the prey that is killed? It turns out it is not the hunter who killed the animal, it is the artisan who made the arrow. It is his kill. But this is merely a matter of bookkeeping, because everyone gets part of the kill, except that the arrowsmith has a right to a favored part. In fact, there is very little in the way of property. They are a nomadic people and can own only what they can carry with them—except for pots and some pieces of clothing and hunting apparatus and things of that sort. And even some of that (there is no personal property) is community property. There is no head man or head woman per se. And there is a cosmology, there is a kind of religion, there is the active encouragement of the religious experience which is obtained, as in many cultures—in fact, all cultures as far as I know—partly by the use of chemical hallucinogens and partly through the use of particular kinds of behavior: dance, trances, and so on. They recognize other levels of consciousness, of conscious experience. They consider these religious experiences or hallucinations as highly valuable, as not something to be laughed at or put into a category of beliefs of the weak-minded. This is a culture in which there has traditionally always been enough to eat. Mainly mongongo nuts, the staple provided by the women, with the men providing the occasional appetizers of meat.

Now, it's interesting to compare such cultures with other cultures that, in a certain sense, because of the biases of our own culture, we know much better. And these are cultures like the Jívaro of the Amazon Valley, in which there are in this world and the next, very striking dominance hierarchies in which there is always someone above someone else, except of course for the Supreme Creator God, above whom there is no one else. These are people who torture their enemies, who do not hug their children—in fact, brutalize their children—who are dedicated to warfare, whose sacrament is not some exotic hallucinogen but instead is ethanol, ordinary ethyl alcohol (I mean, ordinary in our society). And in virtually all the aspects that I just mentioned, there is a completely different way of looking at the world.

Now, these two views—we might call one with a powerful social hierarchy and the other with an almost nonexistent social hierarchy—cut through the anthropological literature. And there's an extremely interesting statistical study by the American social scientist James Prescott, in which he has looked at the compilation by Stanford anthropologist Robert Textor of hundreds of different societies, not all of them still extant. In some cases, for example, from Herodotus, you can get the key characteristics of some society now long dead. And Textor just puts the various categories down as a compilation. What Prescott has done is to do a multivariant analysis, statistical correlation—what goes with what. And the things that apparently go with each other are essentially the two sets of characteristics I just described. It is Prescott's view that there are causal relations. That, in fact, in his view the key distinction has to do with whether cultures hug their children and whether they permit premarital sexual activity among adolescents. In his view those are the keys. And he concludes that all cultures in which the children are hugged and the teenagers can have sex wind up without powerful social hierarchies and everybody's happy. And those cultures in which the children are not permitted to be hugged because of some social ban and a premarital adolescent sexual taboo is strictly enforced wind up killing, hating, and having powerful dominance hierarchies.

Now, you cannot prove a causal sequence from a statistical correlation. And you could just as well argue that what the religious forms are determines everything or what the sacrament is has a powerful connection, between societies with alcohol and the societies that torture their enemies and abuse women and so on. But these correlations at least show that there are two and probably a multiplicity of ways of being human. That these cultures, which as far as we can tell have not been powerfully influenced by Western technical civilization, are yet strikingly different, and the reason for that difference—whatever other reasons there are—must be within us.

And, in fact, if you look at nonhuman primates, you find that some of them have this pecking-order dominance hierarchy and others don't. And it is very likely that built in to humans are both ways of behaving; that is, a hardwired circuit in our brains that permits us to fit effortlessly—or with little effort—into some dominance hierarchy. After all, the military establishments of all nations work, and part of the reason they work is that we must have some predisposition to fit into a dominance hierarchy. And at the same time, we must also have some predisposition for the antithesis, which for short I will call democracy. They lead a kind of uneasy coexistence you can find in any democracy that has a military or a caste system or a class system.

Now, if you grant me that much, let us then go on to the issue of the early function and origins of religion. Clearly there are no observers in our time who were present hundreds of thousands of years ago, and there can be no confident assertions on this subject. All we can have is differing degrees of plausibility. But I think this is, whether you agree with each point I'm making or not, a very useful way to look at the origins of religion. And I'm certainly not the first person to do so. Democritus is quoted as having said in the fifth century B.C.,

> The ancients seeing what happens in the sky, for example, thunder and lightning and thunderbolts and conjunctions of the stars and eclipses of the Sun and Moon were afraid, believing gods to be the cause of these.

This is what is sometimes called "animism," the idea that there are intelligent forces of nature that exist in everything. The Greeks put a minor god in every tree and stream. All of this has been brilliantly discussed by a former Gifford lecturer, Sir James Frazer, in his book *The Golden Bough*. One thing we do if we believe that there is a god of the thunderbolt and do not wish to be hit by a thunderbolt is to propitiate the god of the thunderbolt, to do something to calm him down, to explain that while there may be other targets of thunderbolts deserving of his attention, we are not among them. And we then have to do something to show our respect for him, that we are not talking back to him, that we humble ourselves before him, that we are reverent before him. And many cultures have such institutionalized propitiation, which sometimes goes as far as human sacrifice; that is, to really show you how reverent I am, I will kill what is most dear to me, because you sure couldn't think that I was only playacting if I do that.

The story of God's commandment to Abraham to kill his son, Isaac, is an example of the transition from human to animal sacrifice. After a while people decided it really wasn't worthwhile killing their own children in this way; they would symbolically kill their own children by just getting a goat and killing it. In fact, the general decline in the practice of human and animal sacrifice in the evolution of religion is worth some attention. The Judaic and therefore also the Christian-Islamic religions began when human and animal sacrifice was all the rage.

What does that kind of propitiation mean? It is a wish for the course of nature to be different from what it otherwise would be. It provides the illusion that by some sequence of ritual actions we are able to influence forces of nature that are otherwise inaccessible to us. And therefore it involves a change from the usual course of nature, which was described very nicely by Ivan Turgenev as follows: "Whatever a man prays for, he prays for a miracle. Every prayer reduces itself to this: 'Great God, grant that twice two be not four.'" And from a different tradition, let me quote a Yiddish proverb, which goes, "If praying did any good, they would be hiring men to pray."

Now, does prayer do any good or not? It certainly is still with us. It certainly is connected with those activities of our ancestors, and, as I will argue in a moment, it's certainly connected with the behavior of all of us when we are children. Sir Francis Galton, a cousin of Charles Darwin, said, "Here we've been praying for all these years and nobody seems to know if it does any good or not. Is there a statistical test of the efficacy of prayer?" And he concluded that of course there is. Especially in Britain, because not only do people pray in Britain but people pray differentially. Some people are more in the prayer business than others. Do those who pray more get favors from heaven more? This is in late Victorian times, when these particular views were still more outrageous than they are today. So here is just a little hint of Galton's approach, his sense of scientific protocol:

> There are many common maladies whose course is so thoroughly well understood as to admit of accurate tables of probability being constructed for their duration and result. Such are fractures and amputations. Now, it would be perfectly practicable to select out of the patients at different hospitals under treatment for fractures and amputations two considerable groups. The one consisting of markedly religious and piously befriended individuals, the other of those who were remarkably cold-hearted and neglected. An honest comparison of their respective periods of treatment and the results would manifest a distinct proof of the efficacy of prayer, if it existed to even a minute fraction of the amount that religious teachers exhort us to believe.

And then he goes on to say,

> An enquiry of a somewhat similar nature may be made into the longevity of persons whose lives are prayed for. Also, that of the praying classes generally.

And so he then goes on to compare the mean longevity of sovereigns with that of other classes of persons of equal affluence and gives a table of the results. And the conclusion he states as follows:

The sovereigns are literally the shortest-lived of all
who have the advantage of affluence,

from which he deduces that the efficacy of prayer is not yet demonstrated.

⁘

Now, this has not led to a school of people who do statistical tests of the efficacy of prayer. Hard to know why not. Except that people who don't believe in prayer perhaps are not very interested in this, and those who do are convinced of its efficacy and therefore do not need to go to statistical tests. There is no question that there is something about prayer that seems to work. Surely it provides solace and comfort. It's a way of working through problems. It's a way of reviewing events that have happened, of connecting the past with the future. It does something good. But that doesn't mean that it is as alleged. It doesn't say anything about the existence of a god. It doesn't say anything about the external world. It is a procedure, which on some level makes us feel better.

I maintain that everyone starts out with that sort of attitude. We all grow up in the land of the giants when we are very small and the adults are very large. And then, through a set of slow stages, we grow up, and we become one of the adults. But still within us, surely, is some part of our childhood that hasn't disappeared and hasn't grown up. It's just there. In your formative years, you then learn from direct experience, absolutely incontrovertible, that there are much larger, much older, much wiser, and much more powerful creatures in the universe than you. And your strongest emotional bonds are to them. And, among other things, they are sometimes angry with you, and then you have to work through the anger. And they ask you to do things that you may not want to do, and you must propitiate them, you must apologize, you must do a set of things. Now, how likely is it that after we are all grown up we've fully detached ourselves from this formative experience? Isn't it much more likely that there remains a part of us that is still in the practice of this kind of childhood dealing with parents and other adults? Could that have something to do with prayer specifically and with religious beliefs in general?

Well, this is in fact the scandalous view of Sigmund Freud in *Totem and Taboo* and *The Future of an Illusion,* and other famous books of the first few decades of the twentieth century. And Freud's view was that "at bottom God is nothing more than an exalted father." Of course Freud was living in Vienna at the end of the nineteenth century, in a very patriarchal kind of Judeo-Christian tradition, and therefore it was a very patriarchal kind of god. So it may be that his conclusions do not apply to all religions and all societies, but it's very easy to understand that those religions and those societies lent themselves very much to the Freudian hypothesis.

To say it still more explicitly, the view here is that we start out with the sense that our parents are omnipotent and omniscient, we develop certain relations with them—different degrees of mental health in those relationships, depending on the nature of the relationship between the parents and the child— and then we grow up, and as we do so, we discover that our parents are not perfect. No one is, of course. There is a part of us that is deeply disappointed. There's a part of us that has been inducted into a dominance hierarchy and doesn't like the uncertainty of having to deal with things for ourselves. You know, one of the many reasons that are given for the advantages of military life and other powerfully hierarchical societies is that it's not required to think for oneself very much. There's something calming about that. And so, according to Freud, we then foist upon the cosmos our own emotional predispositions. You may or may not think that this explains a great deal about religion, but it is something I believe worth considering. Fyodor Dostoyevsky wrote in *The Brothers Karamazov*,

> So long as man remains free he strives for nothing
> so incessantly and so painfully as to find someone to
> worship.

I would like now to turn to a related subject, and that has to do with the influence of molecules on the emotions and perceptions. By molecules I just mean chemicals—natural chemicals in the environment or synthetic chemicals made in laboratories. We, of course, all understand that behavior is modified by molecules. Humans all over the world have had experience with substances like ethanol that certainly produced changes in behavior and attitudes and perceptions of the world. We know about tranquilizers that likewise do that. But let us consider a very specific case, and that is manic-depressive syndrome. It's a terrible disease. The manic-depressive swings between two extremes, and it's hard for me to see which is more ghastly: one in the utter pit of despair and the other a kind of high-flying exaltation in which everything seems possible—to the extent that many sufferers of this disease when they are at the manic end of the pendulum believe that they are God. And this is, of course, disabling. Both ends of the swing are disabling, and you don't spend much time in the middle, just like a pendulum, in which you move more slowly at the ends than you do through the middle. It's a disease found in every human culture, and until the last two or three decades there was no effective treatment. Well, there is now a material that powerfully ameliorates manic-depressive syndrome in many patients, provided the dose of this material is administered very carefully. People who have taken this substance in regularly controlled doses, many of them, find that they are able to function again. Their lives are normalized, and they consider it a great blessing. What is this material? It is lithium, a salt. Lithium is a chemical element, the third simplest after hydrogen and helium. It's astonishing that such a simple material could have so profound an effect on a subset of the human population and change not just behavior; if you talk to ex-manic-depressives—that is, manic-depressives whose disease is controlled by regular administration of lithium—their account from the inside of how transforming this treatment is, is really stunning.

Now, bearing this in mind, who will say that there are human emotions that will not, at least one day, be understood in some fundamental manner in the language of molecular biology and neuronal architecture? If you run through our own society and other societies, you find a vast range of substances, many of them chemically very distinct, that powerfully affect mood and emotion and behavior. Not just ethanol but caffeine, mushrooms, amphetamines, tetrahydrocannabinol and the other cannabinoids, lysergic acid diethylamide—known as LSD— barbiturates, Thorazine. It's a very long list.

This prompts certain questions: Are all human emotions to some extent mediated by molecules? If a molecule ingested from the outside can change behavior, is there generally some comparable molecule on the inside that can change behavior? This is now a field that has made remarkable progress. I'm talking about the enkephalins and the endorphins, which are small brain proteins.

In labor, women are amazingly strong in bearing pain, and of course there is a great deal of pain in childbirth. But in that case and in many other traumatic situations, the human body produces a particular molecule that reduces our susceptibility to pain. And it does it for very good survival reasons, which are not hard to understand. There are specific receptors in the brain for these small brain proteins, and it turns out that the opiates ingested from the outside are extremely similar chemically to a particular enkephalin having to do with resistance to pain that is produced on the inside; that is, it is looking as if every time a molecule on the outside does something about human emotions, there is a related molecule on the inside that is naturally produced, which is how it is that we have a brain receptor for this particular kind of molecular functional group.

Let me be a little less abstract and speak from personal experience. I go to the dentist, and he gives me an injection of Adrenalin. It is a molecule. It's a molecule produced in your body, but it's also produced outside. And every time I've had this injection, I'm almost overcome with two contradictory emotions, one of which is to attack the dentist and the other is to leave the dentist's office, both of which I suppose could be understood just on purely rational grounds, considering the circumstances. But this is what adrenaline, the hormone epinephrine, does under any circumstances, under the most benign circumstances. It's called the fight-or-flight syndrome. This molecule makes you either aggressive or, if you want to think about running away, cowardly, one or the other. Very remarkable that two such apparently contradictory emotions can be brought about by the same molecule. But more important than that, it's extremely interesting. They just put this molecule in your bloodstream, and suddenly you feel things. It's just a function of the molecule being there. It's nothing, necessarily, in the external world. And we can understand the reasons for that. Consider our remote ancestors faced with, let us say, a pack of hyenas, not having yet deduced that hyenas with fangs bared are dangerous. It would be too inefficient to have our ancestor consciously stop and think, "Oh, I see those beasts have sharp teeth; they probably can eat somebody. They're coming at me. Maybe I should run away." By then it's too late.

What you need is one quick look at the hyena, and instantly the molecule is produced, and you run away, and later you can figure out what happened. And you can see two populations, one of whom has to slowly think the matter out, the other of whom can rapidly respond to the adrenaline. After a while these guys leave lots of offspring, those guys don't. Everybody winds up generating adrenaline. Natural selection. Not hard to understand how that comes about. And there are, of course, many other molecules like that.

Another one is testosterone, which is produced in males at adolescence and instigates all sorts of bizarre behavior that we all know. I don't want to suggest that at the same age I was immune from it. I personally know the consequences of testosterone poisoning. You might imagine that our distant ancestors could figure out that it was useful to propagate the species and leave offspring and had an intellectual understanding of how it comes about. But this is very iffy. It's requiring a great deal of intellectual activity and cerebration, and it's much better to simply have the whole thing hardwired in the brain and triggered by this molecule after the biological clock has ticked away for a certain period of time. And so the presence of an attractive member of the opposite sex immediately leads to this sequence of events, and the species continues.

There are many other such molecules. Of course, females have estrogen and other hormones. The number of sex hormones is more than one each. Statistics on the subjects that adults of all ages dream about most have sex very high up, and everything else is far below. It's clear the more interested in sex people are, generally speaking, the more offspring they tend to leave, at least before the invention of birth-control devices, and so there is a selective advantage for each species to have this kind of internal machinery.

In just the same ways as the enkephalins and the endorphins and sex hormones influence our sexual activity, what about hormones and religion? People certainly have spontaneous religious experiences. Sometimes they're brought about by deprivation, as with the fasting monks in the desert. There are a number of ways in which sensory deprivation can bring about these experiences. They also happen spontaneously to people in many different cultures, always using the language of the indigenous culture to describe the experience. But also they can be brought about in a molecular way. And certainly the uniform experience, especially in the 1950s and '60s—pioneered by Aldous Huxley and others—was that LSD and other such molecules produce religious experiences. And there were many religionists who objected to this, because they thought it was too easy; that is, you're not supposed to have a religious experience without doing some significant personal deprivation. Just taking, whatever it was, five hundred micrograms of a tablet, was considered too easy.

Let's say there's a molecule that produces a religious experience, whatever the religious experience is. How does that come about? Virtually every time someone takes that molecule, he or she has a religious experience. Does that not suggest that there is a natural molecule that the body produces whose function it is to produce religious experiences, at least on occasion? What could that molecule be like? Let's give it a name, since nobody's discovered it yet, and of course it may not exist—a good one would be "theophilline," but that has already been preempted for an antiasthma drug. And I think "theotoxin" would be biasing the issue too strongly. So let's call it "theophorin," a material that makes you feel religious.

What could the selective advantage of a theophorin be? How would it come about? Why would it be there? Well, what is the nature of the experience? The nature of the experience has, as I say, many different aspects. But one uniform aspect of it is an intense feeling of awe and humility before a power vastly greater than ourselves. And that sounds to me very much like a dominance-hierarchy molecule or part of a suite of molecules whose function it is to fit us into the dominance hierarchies—to suit us for the quest that was, according to Dostoyevsky, to strive for nothing so incessantly and so painfully as to find someone to worship and obey.

Now, what's the good of that? Why would that have any selective advantage? If for no other reason, it would produce social conformity, or, put in more favorable terms, it would ensure social stability and morality. And this is, of course, one of the principal justifications of religion. Any cosmological aspect of the deities is an entirely separate attribute. Consider how we bow our heads in prayer, making a gesture of submission that can be found in many other animals as they defer to the alpha male. We're enjoined in the Bible not to look God in the face, or else we will die instantly. Submissive males of many species, including our own, avert their eyes before the alpha male. In the court of Louis XIV, as the king passed, he was preceded by courtiers crying *"Avertez les yeux!* Avert the eyes! Don't look up. He's passing."* And to this day many animals with a taste for dominance can be made aggressive simply by looking them in the eye. Well, I don't claim that this is the same as all aspects of the religious experience. I think there is as much difference between the religious experience and the bureaucratic religions as there is, say, between sex with love and sex without love. And of course humans have added something profound and beautiful in both cases to the molecular reflex. Perhaps this account will sound tasteless or unpalatable to many, and if so I apologize. But if we treat the question of the origin of religion and the religious experience as a scientific question, then we must ask, "What essential aspects of the religious experience are left out by this hypothesis?" and note that it is at least in principle testable by finding the theophorin, and you could then of course see a large number of controlled experiments to test that out in great detail.

Now, whether or not this explanation is right, there is no question that religions have historically played the role of making people contented with their lot. And it is customary even today to argue that the actual truth or falsity of the religious doctrine does not matter so much as the degree of social stability it brings about. People who through no fault of their own have much less in the way of material goods or respect in a society are told in many religions, "It doesn't matter in this life. Yeah, it looks like you're getting a bad deal, but this is just the twinkling of an eye. What really matters is the next life, and there an implacable cosmic justice awaits you. All those who seem unjustly enriched by the rewards of this life will be punished greatly in the next, whereas you who are the hewers and carriers, the humble people who are content with your lot in this life, will be raised to glory in the next."

Maybe it's true. But it's not hard to see that such a doctrine would be very appealing to the ruling classes of a society. It calms any revolutionary tendencies or even mild complaints and therefore has powerful utility. Many societies, for this reason alone, encourage the contentment with your lot that the religious promise of heaven affords.

Many religions lay out a set of precepts—things people have to do—and claim that these instructions were given by a god or gods. For example, the first code of law by Hammurabi of Babylon, in the second millennium B.C., was handed to him by the god Marduk, or at least so he said. Since there are very few Mardukians today, perhaps no one will be offended if I suggest that this is a bamboozle, that it's a pious hoax. That if Hammurabi had merely said, "Here's what I think everybody ought to do," he would have been much less successful, although he was king of Babylon, than if he said, "God says you should do this."

I recognize that the next step, saying that other lawgivers who are better known today are in the same situation, might produce some degree of outrage at the impiety, but I ask you to nevertheless think it through. Is it not likely that in earlier times, in less sophisticated circumstances, those who wished to impose a certain set of behavioral tenets claimed that they had been handed them by a god or gods.

Now, as soon as you say that religious belief and conventional morality are necessary to keep the society going, you raise the suspicion that these are tools by which those who control the country tend to keep everybody else in line.

And I would like to jump headfirst into a contemporary issue just to make this a little less abstract. Everyone knows about what's going on in apartheid South Africa. I would merely like to draw your attention to something recently produced, called the Kairos Document, derived from a Greek word meaning "the moment of truth." It was written by committed Christians of many races who are opposed to the apartheid system in South Africa. And in the context of what we were just talking about, let me just paraphrase a couple of paragraphs to get a feel about this. It says that state theology in South Africa employs almost exclusively the apostle Paul's view of the state as a power "ordained by God" and commanding obedience. It comes from the remark, "Render unto Caesar what is Caesar's," without there being any detailed explication as to how you go about doing that. The regime elevates the concept of law and order above every other sort of morality.

It goes on to state that

in the present crisis and especially during the State of Emergency, "State Theology" has tried to reestablish the status quo of orderly discrimination, exploitation, and oppression by appealing to the conscience of its citizens in the name of law and order.

And then later on,

This God is an idol. It is as mischievous, sinister and evil as the idols that the prophets of Israel had to contend with. . . . Here we have a God who is historically on the side of the white settlers, who dispossesses black people of their land and gives the major part of the land to his "chosen people." . . . It is the God of teargas, rubber bullets, sjamboks, prison cells and death sentences. Here is a God who exalts the proud and humbles the poor, the very opposite of the God of the Bible. . . .

How rare it is that religions—especially established religions—take the lead in confrontation with the civil authorities when a monstrous injustice is being done. How often it is that the religious authorities take the safe way and temporize or talk about the afterlife or talk about moving slowly or talk about this not being the proper function of religion. And then, on the other side, how often is it that the established religions make authoritative pronouncements on matters of science, matters of fact, matters where they run the desperate risk of being disproved by the next discovery?

This idea was very nicely summed up by Pierre-Simon, the marquis de Laplace, one of the great scientists in the post-Newtonian age, and also a partisan of the French Revolution. In his *System of the World,* in 1796, he said, "Far from us be the dangerous maxim that it is sometimes useful to mislead, to deceive, and enslave mankind to ensure their happiness."

Well, I have tried in this talk to give a further sense of how it is possible in various sorts of ways, ranging from brain chemistry to the wish of the political establishments to maintain power, to understand some of the key aspects of religious belief. By no means does it follow that religions thereby have no function, or no benign function. They can provide in a very significant way, and without any mystical trappings, ethical standards for adults, stories for children, social organization for adolescents, ceremonials and rites of passage, history, literature, music, solace in time of bereavement, continuity with the past, and faith in the future. But there are many other things that they do *not* provide.

I would like to conclude with a quote from Bertrand Russell, from his *Skeptical Essays,* published in 1928. I should warn you, this is redolent with irony.

I wish to propose for the reader's favorable consideration a doctrine which may, I fear, appear wildly paradoxical and subversive. The doctrine in question is this: that it is undesirable to believe a proposition when there is no ground whatever for supposing it true. I must of course admit that if such an opinion became common it would completely transform our social life and our political system. Since both are at present faultless this must weigh against it.

E i g h t

⁂

CRIMES AGAINST CREATION

Tradition is a precious thing, a kind of distillation of tens or hundreds of thousands of generations of humans. It is a gift from our ancestors. But it is essential to remember that tradition is invented by human beings and for perfectly pragmatic purposes. If instead you believe that the traditions are from an exhortatory god and hold that the traditional wisdom is handed down directly from a deity, then we are much scandalized at the idea of challenging the conventions. But when the world is changing very fast, I suggest survival may depend precisely on our ability to change rapidly in the face of changing conditions. We live in precisely such a time.

Consider our past circumstances. Imagine our ancestors, a small, itinerant, nomadic group of hunter-gatherer people. Surely there was change in their lives. The last ice age must have been quite a challenge some ten to twenty thousand years ago. There must have been droughts and new animals suddenly migrating into their area. Of course there is change. But by and large the change is extraordinarily slow. The same traditions for chipping stone to make spears and arrowheads, for example, continues in the East African paleoanthropological sites for tens or hundreds of thousands of years.

In such a society, the external change was slow compared to the human generation time. Back then traditional wisdom, parental prescriptions, were perfectly valid and appropriate for generations. Children growing up of course paid the closest attention to these traditions, because they represented a kind of elixir of the wisdom of previous generations; it was constantly tested, and it constantly worked. It is not for nothing that ancestors were venerated. They were heroes to subsequent generations, because they passed on wisdom that could preserve lives and save them.

Now compare that with another reality, one in which the external changes, social or biological or climatic or whatever we wish, are rapid compared to a human generation time. Then parental wisdom may not be relevant to present circumstances. Then what we ourselves were taught and learned as youngsters may have dubious relevance to the circumstances of the day. Then there is a kind of intergenerational conflict, and that conflict is not restricted to intergenerational but is also intragenerational, internally, because the part of us that was trained twenty years ago, let's say, must be in some conflict with the part of us that is trying to deal with the difficulties of today. So I claim that there are very different ways of thinking for these two circumstances: when change is slow compared to a generation time and when change is fast compared to a generation time. There are different survival strategies. And I would also like to suggest that there has never been a moment in the history of the human species in which so much change has happened as in our time. In fact, it can be argued that in many respects there never will be a time when the change can be so rapid as it has been in our generation.

For example, consider transportation and communication. Just a couple of centuries ago, the fastest practicable means of transportation was horseback. Well, now it is essentially the intercontinental ballistic missile. That is an improvement from tens of miles per hour to tens of miles per second in velocity. It's a very substantial increment. In communication a few centuries ago, except for rarely used semaphore and smoke-signaling systems, the speed of communication was again the speed of the horse. Today the speed of communication is the speed of light, faster than which nothing can go. And that represents a change from tens of miles per hour to 186,000 miles per second. And never will there be any improvement on that velocity.

Now, it's a very different world if the fastest that a message can get to us goes from the speed of a horse or a caravel to the speed of light. The speed of light means that we can talk—in essentially real time—to anybody on the Earth or even on the Moon. Or consider medicine. A few centuries ago, most of the children born to the great houses of Europe died in childhood. And they had the exemplary medical care of the age. Today even quite poor people in some nations at least have infant mortality astonishingly less than the crowned heads of state in the seventeenth century. Or consider the availability of safe and inexpensive means of birth control. It immediately implies a revolution in human relations and especially in the status of women. These are all things that have happened very recently, and you can think of many, many others, all of which involve not just a change in the technical details of our lives but changes in how we think about ourselves in the world. Very major changes, and therefore not a circumstance where the wisdom of, say, the sixth century B.C. is necessarily relevant. It might be, but it might not be. And therefore, for this reason as well—for this reason especially—wisdom may lie not in simply the blind adherence to ancient tenets but in the vigorous and skeptical and creative investigation of a wide variety of alternatives.

For me personally, the kind of science that I do is utterly unthinkable in any other age. I find myself engaged in the spacecraft exploration of nearby worlds, something that would have been considered the most rank fantasy just two generations ago, when the Moon was the paradigm of the unobtainable. Some of you will remember those poems and popular songs—"Fly Me to the Moon," meaning asking for the impossible. And yet in our time a dozen human beings have walked on the surface of the Moon. And as I will stress in tomorrow's talk, that same technology that permits us to travel to other planets and stars also permits us to destroy ourselves—on a global scale, on a scale unprecedented in all of human history, and the mere knowledge that this is possible, even if we are lucky enough for it never to come about, must powerfully influence the lives of everybody who grows up in our time in a way that was not true for any other generation in human history.

I've spent much of my time over the last twenty years in the exploration of the solar system. Our robot emissaries have left the Earth, have visited every planet known to the ancients, from Mercury to Saturn, and reconnoitered some forty attendant smaller worlds, the satellites of those planets. We have flown by all those worlds, we have orbited and landed on three of them: the Moon, Venus, and Mars. There are something approaching a million close-up pictures of other worlds in our libraries. And it is a remarkable experience. Here's a world never before known by human beings, and then, for the first time, it is explored. This is a continuation of the spirit of adventure that I think has been a propelling force in human history. The worlds are lovely. They're exquisite. It is a kind of aesthetic experience to see them.

In the case of Mars, because of the Viking missions, we have been on the surface of that planet for some years, at least in two locales, and have essentially every day examined our surroundings. I personally spent in a certain sense a year on Mars in the course of that mission. I spent at least a great deal of my waking moments thinking about Mars. Now, at the end of such an experience, I feel something I hadn't planned on. And it is that these worlds, as exquisite and instructive as they are, are, as far as we can tell at this point, lifeless. There is in that lovely Martian landscape not a footprint, not an artifact, not even an old beer can, not a blade of grass, not a kangaroo rat, not even, so far as we can tell, a microbe. Mars and the Moon and Venus, as far as we can tell—the only planets we've landed on—are utterly lifeless. Maybe there's life in some places we haven't looked on those worlds. Maybe there used to be life and it is no longer. Maybe there one day will be life. But as far as we can tell here and now, there is none.

After that sort of experience, you then look back on your own world and you begin to have a kind of special feeling for it. You recognize that what we have here is in some sense rare. As I've argued previously, I suspect life and intelligence are a cosmic commonplace. But not so common that they're on every world. And in fact in our solar system we may discover that there is life only on this world.

This says that life is not guaranteed, that life requires something special, something improbable. I'm not for a moment suggesting it requires miraculous, divine, mystical intervention. But in a natural world, you can have probable events and you can have improbable events. And I'm sure this depends on the nature of the environments of the other planets. But there isn't any other planet that's just like the Earth, and, so far as we know so far, there isn't any other planet that has life on it. There are certainly premonitions and stirrings of life, the kind of organic chemistry on Titan, the big moon of Saturn that I referred to earlier. But that's still not the same as life. And so, by performing a first cursory inspection of our solar system, one realizes something important about where we come from.

When you investigate the vistas of time, you find something very similar. Because it is clear from the fossil record that almost every species that has ever existed is extinct; extinction is the rule, survival is the exception. And no species is guaranteed its tenure on this planet. I would like to describe to you one event that I've already referred to as central to the origin of the human species, because it is connected with the main topic of this talk. This is the worldwide extinction event that happened 65 million years ago, at the boundary between the Cretaceous and Tertiary periods of geological time, which also corresponds to the end of the Mesozoic age and the beginning of more recent times.

This is a close-up of a cliff base on a roadside near Gubbio in northern Italy. You can make out the scale of the image from the edge of a five-hundred-lire piece right up at the top. The surface crust has been scraped away a little bit, and the white material is calcium carbonate, essentially chalk, similar to the composition of the White Cliffs of Dover. These are the remains of countless small microorganisms that lived in the Cretaceous seas, forming little calcium carbonate shells that slowly fell through the warm waters of those seas and built up, during

TOP

BOTTOM

fig. 35

Cretaceous time, for many millions of years. This deposit, as you can see, comes to an abrupt end. Time is increasing toward upper left. A layer of reddish brown rock lies above the older white carbonate, separated by a sharp boundary. And it's below this boundary that you find the last dinosaurs, and above the boundary you find an astonishing rate of proliferation of the small mammals into larger mammals, the events that are prerequisite for our own origins.

The sharpness of this boundary worldwide suggests some quite recent catastrophic event. The boundary is that thin layer of gray clay running diagonally across the picture. The clay— this is also true worldwide—has a quite high concentration, an anomalously high concentration, of a chemical element called iridium and other elements like it in the platinum group of metals. It is known that asteroids, and presumably cometary nuclei as well, have much higher abundances of iridium than do ordinary rocks on the Earth. And this iridium anomaly, now supported by a wide range of other data, is generally taken to be evidence for what happened to extinguish the dinosaurs and most of the other species of life on the Earth 65 million years ago.

This is an artist's conception of an object, maybe an asteroid, maybe a cometary nucleus, impacting the Cretaceous oceans. It's about ten kilometers across. It is bigger than the thickness of the ocean, so it is the same as impacting on land. The net result is to carve out in the ocean floor an immense crater and propel the fine particles thus generated into high orbit, making a vast

fig. 36

cloud of pulverized ocean bottom and pulverized impacting object that takes some years to settle out from the Earth's high atmosphere. During that period of time, sunlight is impeded from reaching the surface of the Earth, and the net result is a darkened and cold surface worldwide, which led, because of the differences in mammalian and reptilian physiology, to the extinction of the dinosaurs and many other kinds of life.

That is what happened to the dinosaurs. They were powerless to anticipate it and certainly to prevent it. What I would like now to describe is a catastrophe that in some respects is quite similar, one that endangers the future of our own species. It is very different in one respect: Unlike the dinosaurs, we ourselves, at enormous cost in treasure, have created this danger. We are solely responsible for its existence, and we have the means of preventing it, if we are sufficiently courageous and sufficiently willing to reconsider the conventional wisdom. That problem is nuclear war.

The bombs that destroyed Hiroshima and Nagasaki—everybody's read about them, we know something about what they did—killed some quarter of a million people, making no distinctions according to age, sex, class, occupation, or anything else. The planet Earth today has fifty-five thousand nuclear weapons, almost all of which are more powerful than the bombs that destroyed Hiroshima and Nagasaki and some of which are, each of them, a thousand times more powerful.* Some twenty to twenty-two thousand of these weapons are called strategic weapons, and they are poised for as rapid delivery as possible, essentially halfway across the world to someone else's homeland.

---

*By 2006 the world nuclear arsenals had been reduced to about twenty thousand weapons—still roughly ten times what would be necessary to destory our our global civilization. The principal reductions since 1985 were due to the 1993 Start II Treaty between the United States and the Soviet Union.

The ballistic missiles are sufficiently capable that typical transit times are less than half an hour. Twenty thousand strategic weapons in the world is a very large number. For example, let's ask how many cities there are on the planet Earth. If you define a city as having more than one hundred thousand people in it, there are twenty-three hundred cities on the Earth. So the United States and the Soviet Union could, if they wished, destroy every city on the Earth and have eighteen thousand strategic weapons left over to do something else with.

It is my thesis that it is not only imprudent but foolish to an extreme unprecedented in the events of the human species to have so large an arsenal of weapons of such destructive power simply available. Now, the prompt effects of nuclear war are reasonably well known. I will say a few words about them, but I want to concentrate mainly on the more recently discovered, more poorly known, delayed longer-term and global effects.

Imagine the destruction of New York City by two one-megaton nuclear explosions in a global war. You could choose any other city on the planet, and in a nuclear war you can be reasonably confident that that city would suffer some similar fate. Starting at the World Trade Center and continuing about ten miles in all directions, the effects would play out. You know about the fireball and the shock waves, the prompt neutrons and gamma rays, the fires, the collapsing buildings, the sorts of thing that were responsible for most of the deaths at Hiroshima and Nagasaki. But the bomb light also sets fires, some of which are blown out by the shock wave as the mushroom cloud rises. Others are not.

And these conflagrations can grow. And in many cases, although certainly not all, the conflagrations merge to produce a firestorm. Recent work suggests that firestorms should be much more common and much more severe than had been expected in earlier research, producing the kind of fire as in a well-tended fireplace with an excellent draft. The net result, as advertised: No cities are left standing. But that's the least of the problem.

Beyond the obliteration of the cities is the production of a pall of sooty smoke sitting not just above the city but carried by the fire to quite high altitudes, where this dark smoke is heated by the Sun, which then makes it expand still more. This happens, obviously, not just above one target but above many or most targets.

Cities and petrochemical facilities would be preferentially targeted. Prevailing winds would blow the fine particles in the same direction, from west to east. In anything like a full exchange something like ten thousand nuclear weapons would be detonated.

Some ten days later, there would still be a few nuclear explosions from, I don't know, nuclear-submarine commanders who have not been told that the war was over. The smoke and dust would circulate all around the planet in longitude and spread poleward and equatorward in latitude. The Northern Hemisphere would be almost entirely socked in with smoke and dust. You would see outriders, plumes of smoke in the Southern Hemisphere. The cloud would then cross the equator well into the Southern Hemisphere. And while the effects would be somewhat less in the Southern Hemisphere, sunlight would dim and the temperatures would fall there as well.

Some calculations have been done at the National Center for Atmospheric Research in which a five-thousand-megaton war occurs in July. The widespread distribution of smoke twenty days after the war is over would produce temperature declines as much as fifteen to twenty-five centigrade degrees below normal.

The net result, as you might imagine, is bad. The effects

are global. It appears that they last for months, possibly years. Imagine what disastrous worldwide consequences the destruction of agriculture alone would have. The northern midlatitude target zone is precisely the region that is the principal source of food exports (and experts) to the rest of the world. Even countries nowhere near malnutrition today—Japan, for example—could utterly collapse in a nuclear war from the clouds blown eastward from China, an almost certain target in a nuclear war. Even apart from that, if there were no climatic effects in Japan, and not a single nuclear weapon dropped on Japan, it turns out that more than half the food that people eat there is imported. That alone would kill enormous numbers of people in Japan, and the actual effects would be much worse.

When scientists try to estimate what the consequences of a nuclear war would be, you have to worry not just about the prompt effects. They would be bad enough. The World Health Organization calculates that in an especially nasty nuclear war the prompt effects might kill almost half the people on the planet. You also have to worry about nuclear winter, the cold and the dark that I've just been describing; you have to worry about such facts that those conditions kill not just people and agricultural plants and domesticated animals but the natural ecosystem as well. At just the point that survivors might want to go to the natural ecosystem to live off it, it would be severely stressed.

There is a kind of witches' brew of effects that have been very poorly studied by the various defense establishments, some more than others. These include, for example, pyrotoxins, the smogs of poison gas produced from the burning of modern synthetics in cities, increased ultraviolet light from the partial destruction of the protective ozone layer, and the intermediate timescale radioactive fallout, which turns out to be some ten times more than confident assurances by miscellaneous governments have had it. And so on. The net result of the simultaneous imposition of these independently severe stresses on the environment will certainly be the destruction of our global civilization, including Southern Hemisphere nations, nations far removed from the conflict—nations, if you can find any, that had no part of the quarrel between the United States and the Soviet Union—and, of course, northern midlatitude nations, it goes without saying.

Beyond that, many biologists believe that massive extinctions are likely of plants, of animals, of microorganisms, the possibility of a wholesale restructuring of the kind of life we have on Earth.

It would probably not be as severe as the Cretaceous-Tertiary catastrophe, but possibly approaching it. A number of scientists have said that under those circumstances they cannot exclude the extinction of the human species.

Now, extinction seems to me serious. Hard to think of something more serious, more worthy of our attention, more crying out to be prevented. Extinction is forever. Extinction undoes the human enterprise. Extinction makes pointless the activities of all of our ancestors back those hundreds of thousands or millions of years. Because surely if they struggled for anything, it was for the continuance of our species. And yet the paleontological record is absolutely clear. Most species become extinct. There's nothing that guarantees it won't happen to us. In the ordinary course of events, it might happen to us. Just wait long enough. A million years is quite young for a species. But we are a peculiar species. We have invented the means of our own self-destruction. And it can be argued that we show only modest disinclination to use it.

This is what in a number of Christian theologies is called

crimes against Creation: the massive destruction of beings on the planet, the disruption of the exquisitely balanced ecology that has tortuously grown up through the evolutionary process on this planet. So, since this is clearly recognized as such a theological crime as well as all the other kinds of crimes, it is reasonable to ask where are the religions—the established religions, the incidental independent-thinker religionists—on nuclear war?

It seems to me this is the issue above all others on which religions can be calibrated, can be judged. Because certainly the preservation of life is essential if the religion is to continue. Or anything else. And for me personally, I believe there is simply no more pressing issue. Whatever else we're interested in, it is fundamentally compromised by nuclear war. Whatever personal hopes we have for the future, ambitions for children and grandchildren, generalized expectations for future generations—they are all fundamentally threatened by the danger of nuclear war.

It seems to me that there are many respects in which religions can play a benign, useful, salutary, practical, functional role in the prevention of nuclear war. And there are still other ways that are maybe longer shots but, considering the stakes, are well worth considering. One has to do with perspective.

Now, not all religions have this perspective on the stewardship of the Earth by men and women, but they could. The idea is that this world is not here for us only. It is for all human generations to come. And not just for humans. Or even if you took only a very narrow view of the world, if you were a speciesist in the same sense as being a racist or a sexist, still you would have to be very careful about all those other nonhuman species, because in many intricate ways our lives depend on them. I remind you of the elementary fact that we breathe the waste products of plants and plants breathe the waste products of humans. A very intimate relationship if you think about it. And that relationship is responsible for every breath you take. We in fact depend on the plants, it turns out, a lot more than the plants depend on us. So that sense that this is a world that is worth taking care of is, it seems to me, something that could be at the heart of religions that wished to make a significant contribution to the human future.

Then there are more direct kinds of political activity. For example, religious people played a role in the abolition of slavery in the United States, and elsewhere. Religions played a fundamental role in the independence movement in India and in other countries and the civil rights movement in the United States. Religions and religious leaders have played very important roles in getting the human species out of situations that we should never have gotten into that profoundly compromised our ability to survive, and there is no reason religions could not in the future take on similar roles. There are, of course, occasional circumstances, individual clergypersons who have taken that role in this particular crisis, but it is hard to see any major religion that has made this kind of political activity its foremost objective.

There is also the issue of moral courage. Religions, because they are institutionalized and have many adherents, are able to provide role models, to demonstrate that acts of conscience are creditable, are respectable. They can raise awkward possibilities. The pope, for example, has raised (although not answered) the question about the moral responsibility of workers who develop and produce weapons of mass destruction.

Or is it okay as long as there is a local excuse? Are some excuses better than other excuses? What are the implications for scientists? For corporate executives? For those who invest in such companies? For military personnel? The archbishop of Amarillo has urged workers at a nuclear-weapons facility in his diocese to quit. So far as I know, no one has quit. Religions can remind us of unpopular truths. Religions can speak truth to power. It's a very important function that is often not carried out by all the other sectors of society.

Religions can also speak to their own sectarian eschatologies, especially where they run contrary to human survival. I'm thinking, for example, about the Christian fundamentalist view in the United States that the end of the world is unerringly predicted in the book of Revelation, that the details in the book of Revelation are sufficiently similar to those of a nuclear war that it is the duty of a Christian not to prevent nuclear war. The Christian who does so would be interfering with God's plan. Now, I know I have stated this somewhat more baldly than the advocates of such views, but I believe that is what it comes down to. Christians can play a useful role in providing a steadying hand on people with such eschatologies, because they're very dangerous.

Suppose someone with such a view were in a position of power, and there was a critical decision that had to be made in a moment, and that person had a little sense that maybe this was the fulfillment of biblical prophecy. Maybe he shouldn't make the effort to avoid this, especially if he believed that he himself will be one of the first people to leave the Earth and appear at the right hand of God. He might be interested to see what that would be like. Why slow it down?

Religion has a long history of brilliant creativity in myth and metaphor. This is a field crying out for apposite myth and metaphor. Religions can combat fatalism. They can engender hope. They can clarify our bonds with other human beings all over the planet. They can remind us that we are all in this together. There are many functions that religion can serve in trying to prevent this ultimate catastrophe. Ultimate for us—I want to stress that we're not talking about the elimination of all life on Earth. Doubtless roaches and grass and sulfur-metabolizing worms that live in hot vents in the ocean bottoms would survive nuclear war. It is not the Earth that is at stake, it is not life on Earth that's at stake, it is merely us and all we stand for that is at stake.

Now, along these lines I should also say that at least some religions have specific suggestions on standards of moral behavior that conceivably could be relevant to this problem. (I don't guarantee it; I don't know. The experiment has not been carried out.) And in particular there is the issue of the Golden Rule. Christianity says that you should love your enemy. It certainly doesn't say that you should vaporize his children. But it goes much further than that. It says not just abide your enemy, not just tolerate him, love him.

Well, it's important to ask, what does that mean? Is this just window dressing, or do the Christians mean it?

Christianity also says that redemption is possible. So an anti-Christian would be someone who argues to hate your enemy and that redemption is impossible, that bad people remain forever bad. So I would ask you, which position is better suited to an age of apocalyptic weapons? What do you do if one side does not profess those views and you claim to be a Christian? Must you adopt the views of your adversary or the views advocated by the founder of your religion? You can also ask, which position is uniformly embraced by the nation-states? The answers to those questions are very clear. There is no nation that adopts the Christian position on this issue. Not one. There's 140-some-odd nations on the Earth. As far as I know, not one of them takes a Christian point of view. There may be some perfectly good reasons for that, but it's remarkable that there are nations that take great pride in their Christian tradition that nevertheless do not see any contradiction between that and their attitudes on nuclear war.

By the way, this is not just Christianity. The Golden Rule was uttered by Rabbi Hillel before Jesus, and by the Buddha centuries before Rabbi Hillel. It is involved in many different religions. But for the moment let's talk about Christianity. It seems to me that the admonishment to love our enemy must be something central to Christianity; it's that strong statement of the Golden Rule that sets Christianity apart. There were no qualifying phrases that said, "Love your enemy unless you really don't like him." It says love your enemy. No ifs, ands, or buts. Now, political nonviolence has worked wonders in our time. Mohandas Gandhi and Martin Luther King Jr. achieved extraordinary, and for many people counterintuitive, victories. It might even be a practical, novel, certainly breathtakingly different approach to the nuclear arms race. Maybe not. Maybe it's flawed and hopeless. Maybe the Christian point of view on this issue is inappropriate to the nuclear age. But isn't it interesting that no nation of Christians has adopted it? The Soviet leaders do not profess to be Christians, so if they do not pursue the path of love, they are not inconsistent with their beliefs. But if the leaders of other Western nations profess to be Christian, then what course of action should they be engaged in? Let me stress I don't necessarily advocate such a policy. I don't know if it would work. It may be, as I say, hopelessly naive. But should not those who make conspicuous public displays of their devotion to Christianity follow what is certainly among the central tenets of the faith?

"Do unto others as you would have them do unto you" has a corollary. Others will do unto you as you do unto them. And that encapsulates, among other things, the history of the nuclear arms race. If this can't be done, then I think politicians who are practitioners of such religions ought to confess and admit that they are failed Christians or aspirant Christians but not full-fledged, unqualified, unhyphenated Christians.

I therefore think that the perspective of the Earth in space and time is something with enormous, not just educational but moral and ethical, force. I believe it is lucky for us that this is the time when pictures of the Earth from space are fairly routinely available. We look at them on the evening weather reports and hardly pause to think what an extraordinary item that is. Our planet, the Earth, home, where we come from, seen from space. And when you look at it from space, I think it is immediately clear that it is a fragile, tiny world exquisitely sensitive to the depredations of its inhabitants. It's impossible, I think, not to look at that planet and think that what we are doing is foolish. We are spending a million million dollars every year, worldwide, on armaments. A million million dollars. Think of what you could do with a million million dollars. A visitor from somewhere else—the legendary intelligent extraterrestrial—dipping down to the Earth and inquiring what we are about and finding such prodigies of human inventiveness and such enormous fractions of our wealth devoted not just to the means of war but to the means of massive global destruction—such a being would surely deduce that our prospects are not very good and perhaps go on to some other, more promising world.

When you look at the Earth from space, it is striking. There are no national boundaries visible. They have been put there, like the equator and the Tropic of Cancer and the Tropic of Capricorn, by humans. The planet is real. The life on it is real, and the political separations that have placed the planet in danger are of human manufacture. They have not been handed down from Mount Sinai. All the beings on this little world are mutually dependent. It's like living in a lifeboat. We breathe the air that Russians have breathed, and Zambians and Tasmanians and people all over the planet. Whatever the causes that divide us, as I said before, it is clear that the Earth will be here a thousand or a million years from now. The question, the key question, the central question—in a certain sense the only question—is, will we?

Nine

THE SEARCH

Without knowing what I am and why I am here, life is impossible.

• Leo Tolstoy, *Anna Karenina* •

If we don't find life literally impossible without answering
that question, at least its difficulties increase. It is very reasonable for humans to want to understand something of our
context in a broader universe, awesome and vast. It is also reasonable for us to want to understand something about ourselves.
Since we have powerful unconscious processes, this means that
there are parts of our selves that are hidden from us. And this
two-pronged investigation into the nature of the world and the
nature of our selves is, to a very major degree, I believe, what
the human enterprise is about.

Our success as a species is surely due to our intelligence, not
primarily to our emotions, because many, many different species
of animals surely have emotions. Many, many different species
of animals also have varying degrees of intelligence. But it is
our intelligence—our interest in figuring things out, our ability
to do so, coupled with our manipulative abilities, our engineering talents—that is responsible for our success. Because surely
we are not faster than all other species, or better camouflaged, or

better diggers or swimmers or fliers. We are only smarter. And, at least until the invention of weapons of mass destruction, this intelligence has led to the steady—in fact exponential—increase in our numbers. And in the last few thousand years, our numbers on this planet have increased by much more than a factor of a hundred. There are human outposts not just everywhere on the planet, including Antarctica, but in the ocean depths and in near-Earth orbit. And it is clear that if we do not destroy ourselves, we will continue this progressive, outward movement until there will be human settlements on neighboring worlds.

It seems to me also clear that historians of a thousand years from now, if there are any, will look back on our time as being absolutely critical, a turning point, a branch point in human history. Because if we survive, then this time will be remembered as the time when we could have destroyed ourselves and came to our senses and did not. It will also be the time in which the planet was bound up. And it will also be remembered as the time when, slowly, tentatively, haltingly, we first sent our robot emissaries and then ourselves to neighboring worlds.

Now, all of these are extraordinary and unprecedented activities. Never before have we had the capability of destroying ourselves, and therefore never before have we had the ethical and moral responsibility not to do so. A way of looking at the time we happen to inhabit is as follows: We started hundreds of thousands to millions of years ago as itinerant tribespersons, in which the fundamental loyalty was to a very small group, by contemporary standards. Typical hunter-gatherer groups are maybe a hundred people, so the typical person on the planet had an allegiance to a group of no more than a hundred or a few hundred people.

The names that many of these tribes give to themselves are

touching in their narrowness. All over the world, people call themselves "the people," "the men," "the humans." And all those other tribes, they aren't people, they aren't men, they aren't humans. They are something else. Now, that doesn't mean that a state of constant warfare existed among these tribes, as Thomas Hobbes, for example, imagined. A significant fraction of those early groups, there is reason to think, were benign, calm, peace-loving, not interested in systematic, bureaucratized aggression, which is the function of states at a later time.

As time passed, groups have merged, sometimes voluntarily, sometimes involuntarily, and the unit to which personal identi-fication and loyalties are due has grown. The sequence is known to all of those who take courses in the history of civilization at universities, in which we pass through allegiances to larger groups, to city-states, to settled nations, to empires. Today the typical person on the Earth is obviously a patchwork quilt of political, economic, ethnic, and religious identifications, owing allegiance to a group or groups consisting of a hundred million people or more. It's clear that there is a steady trend, if the trend contin-ues, there will be a time, probably not so far in the future, when the average person's typical identification is with the human species, with everyone on Earth.

The more we view the Earth from the outside, the more we come to see it as an exquisite, tiny world, everyone dependent upon everyone else, the sooner that general perception will come into being. Despite all the faults of international organiza-tions, it is nevertheless striking in our time, in this century and the last few, but especially in this century, that organizations of global purview, involving essentially every nation on Earth, have grown up, have persisted, and we would, of course, not ex-pect them to be perfect. Their imperfections are a function of the newness of the organization and the fact that human beings are imperfect. But it is a trend, a token, of the direction in which we are headed, provided we do not destroy ourselves.

One way to think of our time is as a race between these conflicting tendencies: one to bind up the planet, preserving, it may be, some of its ethnic and cultural diversity, and the contrary trend to destroy the planet, not in the geophysical sense but the planet in the sense of the world that we know. It is by no means clear which of these two conflicting tendencies will win out, in the lifetime of you who are among the first to be hearing these words.

Now, another way of looking at this is as a conflict within the human heart, as a conflict between the bureaucratic, hierarchical, aggressive parts of our nature, which in a neurophysiological sense we share with our reptilian ancestors, and the other parts of our nature, the generalized capacity for love, for compassion, for identification with others who may superficially not look or talk or act or dress exactly like us, the ability to figure the world out that is focused and concentrated in our cerebral cortex. Our survival is (how could we have imagined it to be anything else?) a reflection of our own nature and how we manage these contending tendencies within the human heart and mind.

Since the times are so extraordinary, since they are unprecedented, it is in no way clear that ancient prescriptions retain perfect validity today. That means that we must have a willingness to consider a wide variety of new alternatives, some of which have never been thought of before, others of which have, but have been summarily rejected by one culture or another. We run the danger of fighting to the death on ideological pretexts.

We kill each other, or threaten to kill each other, in part, I think, because we are afraid we might not ourselves know the truth, that someone else with a different doctrine might have a closer approximation to the truth. Our history is in part a battle to the death of inadequate myths. If I can't convince you, I must kill you. That will change your mind. You are a threat to my version of the truth, especially the truth about who I am and what my nature is. The thought that I may have dedicated my life to a lie, that I might have accepted a conventional wisdom that no longer, if it ever did, corresponds to the external reality, that is a very painful realization. I will tend to resist it to the last. I will go to almost any lengths to prevent myself from seeing that the worldview that I have dedicated my life to is inadequate. I'm putting this in personal terms so that I don't say "you," so that I'm not accusing anyone of an attitude, but you understand that this is not a mea culpa; I'm trying to describe a psychological dynamic that I think exists, and it's important and worrisome.

Instead of this, what we need is a honing of the skills of explication, of dialogue, of what used to be called logic and rhetoric and what used to be essential to every college education, a honing of the skills of compassion, which, just like intellectual abilities, need practice to be perfected. If we are to understand another's belief, then we must also understand the deficiencies and inadequacies of our own. And those deficiencies and inadequacies are very major. This is true whichever political or ideological or ethnic or cultural tradition we come from. In a complex universe, in a society undergoing unprecedented change, how can we find the truth if we are not willing to question everything and to give a fair hearing to everything? There is a worldwide closed-mindedness that imperils the species. It was always with us, but the risks weren't as grave, because weapons of mass destruction were not then available.

We have Ten Commandments in the West. Why is there no commandment exhorting us to learn? "Thou shalt understand the world. Figure things out." There's nothing like that. And very few religions urge us to enhance our understanding of the natural world. I think it is striking how poorly religions, by and large, have accommodated to the astonishing truths that have emerged in the last few centuries.

Let's think together for a moment about the prevailing scientific wisdom on where we come from. The idea that nearly 15,000 million years ago the universe, or at least its present incarnation, was formed in the big bang; that for some 5,000 million years thereafter even the Milky Way Galaxy was not formed; that for some 5,000 million years after that, the Sun and the planets and the Earth were not formed; that 5,000 million years ago, on an Earth not identical by any means to the one we know today, a large-scale production of complex organic molecules occurred that led to a molecular system capable of self-replication, and therefore began the long, tortuous, and exquisitely beautiful evolutionary sequence that led from those first organisms, barely able to make vague copies of themselves, to the magnificent diversity and subtlety of life that graces our small planet today.

And we have grown up on this planet, trapped, in a certain sense, on it, not knowing of the existence of anything else beyond our immediate surroundings, having to figure the world out for ourselves. What a courageous and difficult enterprise, building, generation after generation, on what has been learned in the past; questioning the conventional wisdom; being willing, sometimes at great personal risk, to challenge the prevailing wisdom and gradually, slowly emerging from this torment, a well-based, in many senses predictive, quantitative understanding of the nature of the world around us. Not, by any means, understanding every aspect of that world but gradually, through successive approximations, understanding more and more. We face a difficult and uncertain future, and it seems to me it requires all of those talents that have been honed by our evolution and our history, if we are to survive.

One thing that seems especially striking in contemporary culture is how few benign visions of the immediate future are offered up. The mass media show all sorts of apocalyptic scenarios, ghastly futures. And there tends to be a kind of self-fulfilling prophecy to these prognostications. How rarely is it that we see a projection twenty or fifty or a hundred years into the future into a world in which we have come to our senses, in which we have figured things out? We can do that. There's nothing that says that we will inevitably fail to meet these challenges. We have solved more difficult problems, and many times. For example, there was once a doctrine called the divine right of kings. It held that God gave kings and queens the right to rule their people. And at that time it really meant rule. "Rule" was not so very different from "own." And eminent clergymen argued that this was clearly written in the Bible. It was the will of God. Eminent secular theologians, Thomas Hobbes, for example, argued just the same thing. And yet there was a stirring sequence of worldwide revolutions—the American, the French, the Russian, and a number of others—that have now produced a planet in which no one, except an occasional atavistic emperor of a short-lived, small country, no one believes in the divine right of kings. It's now a kind of embarrassment. It's something that our ancestors believed but we in this more enlightened time do not.

Or consider chattel slavery, which Aristotle argued was intended, it was in the natural order of things, the gods required it, that any movement to free the slaves was against divine intention. And slaveholders throughout history have pointed to passages in the Bible to justify the holding of slaves. Yet today, in another stirring sequence of events worldwide, legal chattel slavery has been essentially eliminated. And again it is something from our past that we are embarrassed about, that we surely should still think of as an important insight into a dark side of human nature that should be resisted. Surely the depredations visited on peoples who were once enslaved have not been balanced, but we have made remarkable progress.

Or look at the status of women, about which finally the planet is coming to its senses in our own time. Or even things like smallpox and other disfiguring and fatal diseases, diseases of children, that were once thought to be an inevitable, God-given part of life. The clergy argued, and some still do, that those diseases were sent by God as a scourge for mankind. Now there are no more cases of smallpox on the planet. For a few tens of millions of dollars and the efforts of physicians from a hundred countries, coordinated by the World Health Organization, smallpox has been removed from the planet Earth.

The vested interests in favor of the divine right of kings, or slavery, were very large. Kings had a vested interest in the divine right of kings. Slaveholders had a vested interest in the continuation of the institution of slavery. Who has a vested interest in the prospects of nuclear war? It's a very different situation. Everyone is vulnerable today. And therefore I think it's important to remember that we have dealt with and solved much more difficult problems than this.

The only problem is that the threat of nuclear war has to be dealt with swiftly, because the stakes are too high. The clock is ticking. We cannot permit a leisurely pace.

Suppose you are a linguist. You are interested in the nature and evolution of language. But unfortunately you know only one language. No matter how clever you are, no matter how complete your dictionary of whatever the language is—say,

Nahuatl—you will be fundamentally limited in your ability to generate a broad, interdisciplinary, predictive theory of language. How could you be expected to do very well if you knew only one language? If Newton were restricted, in working through the theory of gravitation, to apples and forbidden to look at the motion of the Moon or the Earth, it is clear he would not have made much progress. It is precisely being able to look at the effects down here, look at the effects up there, comparing the two, which permits, encourages, the development of a broad and general theory. If we are stuck on one planet, if we know only this planet, then we are extremely limited in our understanding even of this planet. If we know only one kind of life, we are extremely limited in our understanding even of that kind of life. If we know only one kind of intelligence, we are extremely limited in knowing even that kind of intelligence. But seeking out our counterparts elsewhere, broadening our perspective, even if we do not find what we are looking for, gives us a framework in which to understand ourselves far better.

I think if we ever reach the point where we think we thoroughly understand who we are and where we came from, we will have failed. I think this search does not lead to a complacent satisfaction that we know the answer, not an arrogant sense that the answer is before us and we need do only one more experiment to find it out. It goes with a courageous intent to greet the universe as it really is, not to foist our emotional predispositions on it but to courageously accept what our explorations tell us.

SELECTED Q & A

A fter each lecture there was a lively question-and-answer period. Unfortunately, the transcripts report that in some cases the audience was not provided with working microphones. These are the fragments of the sessions that survive.

CHAPTER ONE

Questioner: When will we be likely to make contact with another intelligence?

CS: Prophecy is a lost art. But what I would say is that it's clear that if we don't try to seek such intelligence, it will be more difficult to find it. And it is remarkable that we live in a time when the technology permits us, at least in a halting way, to seek such intelligences, mainly by constructing large radio telescopes to listen for signals being sent to us—radio signals—by civilizations on planets of other stars.

Questioner: Considering the accomplishments of scientists like Newton and Kepler, is it likely that science will one day come upon a demonstration of the existence of God?

CS: The answer depends very much on what we mean by God. The word "god" is used to cover a vast multitude of mutually exclusive ideas. And the distinctions are, I believe in some cases, intentionally fuzzed so that no one will be offended that people are not talking about *their* god.

But let me give a sense of two poles of the definition of God. One is the view of, say, Spinoza or Einstein, which is more or less God as the sum total of the laws of physics. Now, it would be foolish to deny that there are laws of physics. If that's what we mean by God, then surely God exists. All we have to do is watch the apples drop.

Newtonian gravitation works throughout the entire universe. We could have imagined a universe in which the laws of nature were restricted to only a small portion of space or time. That does not seem to be the case. And Newtonian gravitation is one example, but quantum mechanics is another. We can look at the spectra of distant galaxies and see that the same laws of quantum mechanics apply there as here. So that is itself a deep and extraordinary fact: that the laws of nature exist and that they are the same everywhere. So if that is what you mean by God, then I would say that we already have excellent evidence that God exists.

But now take the opposite pole: the concept of God as an outsize male with a long white beard, sitting in a throne in the sky and tallying the fall of every sparrow. Now, for *that* kind of god I maintain there is no evidence. And while I'm open to suggestions of evidence for that kind of god, I personally am dubious that there will be powerful evidence for such a god not only in the near future but even in the distant future. And the two examples I've given you are hardly the full range of ideas that people mean when they use the word "god."

. . .

CS: The questioner asked whether I was familiar with Democritus, bearing in mind my suggestion that we now know things that were not known in the past. Democritus is one of my heroes. I think I know more than Democritus. Now, I don't claim to be smarter than Democritus, but I have the advantage that Democritus did not of having twenty-five hundred years of scientists between him and me. So, for example, I'll give you a few things that I know and that Democritus did not know. Democritus proposed that the Milky Way Galaxy was composed of stars. Far ahead of his time. He did not know that there were other galaxies. We know that.

We know of the existence of many more planets than he did. We have examined them close up. We know what their physical natures are. He did not, although he speculated that they were at least made of matter. We have an idea of how many stars there are in the Milky Way Galaxy.

Democritus was an atomist. You will not exceed me in your admiration for Democritus. And were the vision of Democritus to have been adopted by Western civilization, instead of being cast aside for the pale views of Plato and Aristotle, we would be vastly further ahead today, in my personal view.

. . .

CS: The questioner asks have I not perhaps been looking through the wrong end of the telescope; that is, is not the proper province of religion the human heart and mind and ethical questions and so on, and not the universe?

Well, I couldn't agree with you more, except that it is striking how many religions have felt that astronomy is their province and have made confident statements about matters astronomical. It is possible to design religions that are incapable of disproof. All they have to do is to make statements that cannot be validated or falsified. And some religions have very neatly positioned themselves in that respect. Now, that means that you cannot make any statements on how old the world is; you cannot make any statements about evolution; you cannot make any statements about the shape of the Earth (the Bible is quite clear about the Earth being flat, for example), and so on. And then you have religions that are making statements on human behavior, where religions have, in my view, made significant contributions. But it is a very rare religion that avoids the temptation to make pronouncements on matters astronomical and physical and biological.

Questioner: Do you think humans at this time could cope with us finding extraterrestrial intelligence?

CS: Sure. Why not? Well, there's no question that the discovery of something very different will worry people precisely because it's different. Look at the degree of xenophobia in human cultures in which it is other humans, trivially different from us, who are the object of great fear and concern and violence and aggression and murder and terrible crimes. So there's no question that were we to receive a signal, much less come face-to-face, or whatever the appropriate bodily part is, with another intelligent being, there would be a sense of fear, horror, loathing, avoidance, and so on.

But the receipt of a message is a very different story. You are not even obligated to decode. If you find it offensive, you can ignore it. And there is a kind of providential quarantine between the stars, with very long transit times even at the speed of light, that I think obviates, if not altogether eliminates, this difficulty.

. . .

CS: The questioner asks that is not one central goal of religions the idea of a personal god, of a purpose for individuals and for the species as a whole, and is that not one of the reasons for the success on an emotional level (I'm paraphrasing) of many religions? And he then goes on to say that he, himself, does not see much evidence in the astronomical universe for a purpose.

I tend very much to agree with you, but I would say that purpose is not imposed from the outside; it is generated from the inside. We *make* our purpose. And there is a kind of dereliction of duty of us humans when we say that the purpose is to be imposed on the outside or found in some book written thousands of years ago. We live in a very different world than we lived in thousands of years ago. There is no question that we have many obligations to guarantee our purposes, one of which is to survive. And *that* we have to work out for ourselves.

CHAPTER TWO

Questioner: What is your opinion on the nature of the origins of intelligent life in the universe?

CS: I'm for it!

CHAPTER FOUR

Questioner: I'm a wee bit skeptical at Drake's equation. It doesn't really indicate how much extraterrestrial life there is. All it indicates is whether the user of it is a pessimist or an optimist. And given this, why do you bother to use it at all?

CS: That's a perfectly good question. And it has a perfectly good answer. And that is, it *might* have turned out before you went through this exercise that even in the optimistic case the number of civilizations was so low that it didn't make sense to search. But it doesn't turn out that way. There's a sequence of perfectly plausible numbers that lead to a large number of civilizations. It doesn't say it's guaranteed, but it survives the initial test. That's the only function that this has, apart from the very nice fact that there is a single equation that connects stellar astrophysics, solar-system cosmogony, ecology, biochemistry, anthropology, archaeology, history, politics, and abnormal psychology.

Questioner: Oh, this scares the hell out of me. But there's one fact that I think Professor Sagan hasn't brought into account in Drake's formulation. The point is that he's only taken *this* galaxy into account and not all the other—I don't know— thousands or millions of other galaxies, way back to the big bang 15,000 million years ago. So, I mean, if you're going to take that particular formula, why don't you multiply it by that particular factor?

CS: Again, a good question, and I was merely talking about the justification for the search for signals from advanced civilizations in our galaxy. Clearly you can imagine them in some other galaxy. For their signals to reach us here, they have to have a technology far in advance of ours, but that's perfectly possible. And in fact Frank Drake and I have made a search of just a few nearby galaxies with exactly that idea in mind. We found nothing at the few frequencies we looked at. But, you see, once you start imagining signals coming from another galaxy, then you are into significant power levels and therefore significant dedication by some other civilization to try to make contact with what for them would be a distant galaxy. If you imagine civilizations in our own galaxy, you can at least contemplate that they know that this solar system is a plausible abode for life, even if they haven't visited here to check it out, that there's some way that they could target our particular region of the galaxy for a specific message. There's no way that this could be the case from a distant galaxy, as far as I can see.

This does remind me, though, that I forgot to say something. *Very* nearby civilizations *can* detect our presence, and that is because television gets out. Not just television but radar. Radar and television get out. Most of AM radio, for example, doesn't. So let's just look at the television for a moment. Large-scale commercial television broadcasting on Earth begins when? In the late 1940s, mainly in the United States.

So forty years ago there's a spherical wave of radio signals that spreads out at the speed of light, getting bigger and bigger as time goes on. Every year later it's an additional light-year away from the Earth. Now, let's say it's forty years later, so that expanding spherical wave front is forty light-years from the Earth, containing the harbingers of a civilization newly arrived in the galaxy. And I don't know if you know about 1940s television in the United States, but it would contain Howdy Doody and Milton Berle and the Army-McCarthy Hearings and other signs of high intelligence on the planet Earth. So I'm sometimes asked, if there are so many intelligent beings in space, why haven't they come here? Now you know. It's a sign of their intelligence that they haven't come. (I'm just joking.) But it's a sobering fact that our mainly mindless television transmissions are our principal emissaries to the stars. There is an aspect of self-knowledge that this implies that I think would be very good for us to come to grips with.

CHAPTER FIVE

Questioner: How do you recognize the truth when it is upon us?

CS: A simple question: How can we recognize the truth? It is, of course, difficult. But there are a few simple rules. The truth ought to be logically consistent. It should not contradict itself; that is, there are some logical criteria. It ought to be consistent with what else we know. That is an additional way in which miracles run into trouble. We know a great many things—a tiny fraction, to be sure, of the universe, a pitifully tiny fraction. But nevertheless some things we know with quite high reliability. So where we are asking about the truth, we ought to be sure that it's not inconsistent with what else we know. We should also pay attention to how badly we want to believe a given contention. The more badly we want to believe it, the more skeptical we have to be. It involves a kind of courageous self-discipline. Nobody says it's easy. I think those three principles at least will winnow out a fair amount of chaff. It doesn't guarantee that what remains will be true, but at least it will significantly diminish the field of discourse.

Questioner: Have you any comments to make on the Shroud of Turin?

CS: The Shroud of Turin is almost certainly a pious hoax; that is, not a contemporary hoax but a hoax from the fourteenth century, when there was significant traffic in pious hoaxes. And my technical knowledge of the Shroud of Turin comes from Dr. [Walter] McCrone of Chicago, who has worked on it for some years. He found the "blood" to be iron oxide pigments, and there is nothing that cannot be explained by the technology available in the fourteenth century. By the way, there is no provenance of the Shroud of Turin earlier than the fourteenth century.* So I'm sorry that my knowledge is secondhand on this issue, and I know that there are people who believe, for reasons

---

*In 1988 the Vatican allowed samples of the original shroud material to be dated by the radiocarbon method. Three laboratories (in Arizona, Oxford, and Zurich) independently determined that the fabric dates from the period A.D. 1260 to 1390.

that are apparent. No, I'm sorry. I haven't said that fairly. There are people who believe that it is the authentic death shroud of Jesus on the cross. But the evidence is very meager.

Questioner: The religionists proffer ghosts and miracles. The physicists propose equations. What is the fundamental difference between them?

CS: A very good question. How can we tell what's what? One thing we can do is we can check out the explanation in terms of repeatability. Verifiability. So, for example, if physicists after Isaac Newton say that the distance that a falling object falls in time t is a constant times t^2, and if you are skeptical or dubious about that, you can perform the experiment, and you will find that if it takes twice as long to fall, it goes four times farther, and so on. They will also say that the velocity increases proportionately to the time. You can check that. You can drop boulders off bridges, if it's permitted by the local police, and check out these contentions. After a while you get a sense that, at least in this limited realm, the physicists know what they're talking about. What is more, it is remarkable that Buddhist physicists find just the same regularity. And Hindu physicists, and atheist physicists, and Christian physicists, and so on. All find the same laws of nature. Somehow it doesn't depend on the local culture, on the local training. What the physicists say seems to be true all over the Earth. And then you look at other planets. Other stars. Other galaxies. And the same laws apply everywhere.

Now, this doesn't say that every contention of every physicist has this wonderful degree of regularity. Physicists make mistakes just like anyone else. But the way in which physicists have an advantage is that there is a tradition of skepticism and a tradition of mutually checking out each other's contentions. Whereas in religion there is a practice of great reluctance to challenge what any other member of the professional caste says. That is not true in physics. A physicist is almost as delighted in disproving another physicist's contention as in demonstrating some new principle of physics. And you know Newton's famous remark that if he had seen further it was by standing on the shoulders of giants. What he meant was that there is a continuous progress in science. And through this progression of insights, through this mutual checking, the subject advances mightily. Whereas if you take supposed religious proofs of the existence of God, it is really quite remarkable that no new proof has been offered—never mind the validity—no fundamentally new proof has been offered in centuries. The anthropic principle that I talked about in an earlier lecture is as close as you can come, but it is merely a variant on the argument from design.

So I see methodologically a significant difference between how science proceeds and how religion proceeds. Now, an earlier questioner gave a very good example. He said, "Scientists talk about the expanding universe. What began the expansion?" Now, many astrophysicists would say that's not their problem. Their problem is to tell you what the universe is doing but not to tell you *why* it's doing it. They avoid that "why" question— and it's not due to modesty, although it's sometimes phrased in a way to suggest that we don't want to mess around with the really big questions. But physicists love to mess around with the big questions. The reason that questions such as "Why did the universe expand?" are considered off-limits is that there's no experiment you can do to check it out.

. . .

CS: The question has to do with the Bermuda Triangle. This is certainly not significantly different from UFOs and ancient astronauts. It is as good an example. Here is a case where if you track the mysterious disappearances or sinkings of airplanes and ships, you find, it is alleged, a concentration of these disappearances in a triangular region off Bermuda. And the explanations that have been proposed are many, one of which is that there is a UFO on the Atlantic floor that eats airplanes and boats.

Now, there are several things that might be said about this. Is the statistical evidence as purported? In fact, is there *any* statistical evidence? Do we compare? Do the proponents of the Bermuda Triangle "mystery" compare the rate of loss of ships and airplanes off Bermuda to the rate of loss of ships and airplanes in some other region of the world with comparable weather and of equal area and traffic frequency? Nowhere do they attempt that. But others have, and found not a smidgen of evidence that the disappearance rate is larger there than elsewhere.

And also I would raise a related question. Why is it that there are no examples of mysterious disappearances of trains? Train sets out from one station, everything looks fine, and then it is supposed to appear at another station. It's not there. They go back to search along the tracks; it's totally disappeared! The thing about the ocean is you can *sink* in it. It has a natural explanation built in for mysterious disappearances, whereas railroad beds provide awkward opportunities for mysterious disappearances.

There is a famous case that I'll tell and then end. An enormous electrical rotor for a power-generating plant was completed—I've forgotten exactly where this was; let us say in Michigan—to be transported a thousand miles or so on a railway flatbed with the rotor tied down but in a vertical position. It left the factory perfectly all right. The train did arrive at its destination, but with no rotor. Rotor gone. And so, it being a very expensive piece of machinery, the railway detectives (you can imagine this as a change from the usual sorts of cases they have to deal with) go in a small railroad car along every inch of the thousand miles, and there isn't any rotor sitting by the side of the railway bed. So it has disappeared. Supernatural. And insurance companies are involved because it's expensive, so there's a second search. They can't find it. Nobody on the train saw anything amiss.

Twenty years pass, and then about three miles from the railway track a swamp is drained for a housing project, and there, at the bottom of the swamp, is this rotor, which must have broken its moorings and rolled three miles to the swamp. Can you imagine being out for a midnight walk and seeing this apparition rolling by? If anyone had seen it, it surely would have been an impetus to found a new religion.

CHAPTER SIX

Questioner: Well, I'd just like to ask you about your closing remarks. You were talking about possible proofs that God could have left us of His own existence. You don't think that you're making a rather arrogant assumption in that you are assuming that, for example, it could be possible that He has . . . that God has left in these religious writings the types of statements that you are suggesting, but it was simply that we ourselves have not got to that stage of development. For example, if He'd made statements about special relativity, a hundred years ago those would have been still meaningless. Could there not now be statements that in a hundred years would make sense to us that would not make sense to us now? Secondly, a more specific example, some people at the Hebrew University at Tel Aviv claim that there are in the Torah in Hebrew various words or messages in which were concealed the names of some thirty trees in Hebrew, with the letters of each tree equally spaced within the passages. And their suggestion is that it would have been impossible for anyone, without the use of computers, to have devised such complicated messages.

CS: This is from the Kabbalistic tradition?

Questioner: Uh-huh.

CS: I have looked at it a little bit, and I believe it is an example of the statistical error of the enumeration of favorable circumstances; that is—what's the best way to put it?—there is a stunning correlation between earthquakes in the Andes and oppositions of the planet Uranus. Is this a causal connection or not? First thing you ask is, how many connections had to be looked for before this particular one was derived? Volcanoes in Sicily with oppositions of the planet Mars—think of how many volcanoes there are in the world, how many earthquakes there are, how many planets there are, how many stars. If you start making a specific number of cross-correlations you will, of course, on occasion, come upon a coincidence. And what you have to do in a posteriori knowledge is to add up all those other cases of possible coincidences that you looked at or could have looked at.

Now, the cases that you are mentioning seem to me highly ambiguous. And I would ask, among other things, why these results have not been submitted to the leading scientific journals, *Nature*, for example, in Britain, *Science* in America. What kind of peer review have they got? Also, why something so obscure as the kinds of trees? Why not the detailed structure of a thousand amino acid proteins?

On the first part of your question about might there not be such clues waiting for us but we are not smart enough to recognize them: Well, maybe. You could never exclude that. But that is a slim reed upon which to base a religious faith. When they are discovered, *then* let's talk about them, but not until then.

Maybe there is a complete description of everything we want to know lying about on the surface of Pluto. And we won't be there until the middle twenty-first century, so we'll just have to hang on till then. Perfectly okay. Let's talk about it in the middle of the twenty-first century. For now there is no such evidence.

Questioner: In reality He is there. God is love.

CS: Well, if we say that the definition of God is reality, or the definition of God is love, I have no quarrel with the existence of reality or the existence of love. In fact, I'm in favor of both of them. However, it does not follow that God defined in that way has anything to do with the creation of the world or of any events in human history. It does not follow that there's anything that is omnipotent or omniscient and so on about God defined in such a manner. So all I'm saying is, we must look at the logical consistency of the various definitions. If you say God is love, clearly love exists in the world. But love is not the only thing that exists in the world. The idea that love dominates everything else, I deeply hope is true, but there are arguments that can very well be proposed, from a mere glance at the daily newspapers, to suggest that love is not in the ascendant in contemporary political affairs. And I don't see that it helps to say, forgive me, that God is love, because there are all those other definitions of God, that mean quite different things. If we muddle up all the definitions of God, then it's very confusing what's being talked about. There is a great opportunity for error in that case. So my proposal is that we call reality "reality," that we call love "love," and not call either of them God, which has, while an enormous number of other meanings, not exactly those meanings.

Questioner: Dr. Sagan, when you spoke to us yesterday, you mentioned something about Russia's approach to the recording of their history, and you said that Trotsky had virtually been written out of it. And how would you view the case for a corollary to that: Perhaps people can be written *into* history. For example, Jesus Christ?

CS: It's certainly possible. The only evidence for the existence of Jesus is the four Gospels and the subsequent books. And apart from that, there is merely the account of Josephus in the *History of the Jews*, which internal evidence suggests was put in by Christian apologists at a later time. On the other hand, for me personally, I find the accounts in the Gospels reasonably internally consistent, and I don't see any particular problem about Jesus as a historical figure in the same sense of Mohammed and Moses and Buddha. For all of them, I would think the least unsatisfactory hypothesis is that they were real people, genuine historical figures, great men, the details of whose lives and missions have been, of course, distorted by subsequent advocates and enemies both. It's inevitable. It's the way humans go about things.

Questioner: I'd like to ask you about why you think any omnipotent being would want to leave evidence for us.

CS: I think I entirely agree with what you say. There is no reason I should expect an omnipotent being to leave evidence of His existence, except that the Gifford Lectures are supposed to be *about* that evidence. And I hope it is clear that the fact that I do not see evidence of such a God's existence does not mean that I then derive from that fact that I know that God does not exist.

That's quite a different remark. *Absence of evidence is not evidence of absence.* Neither is it evidence of presence. And this is again a situation where our tolerance for ambiguity is required.

The only thrust of these remarks is for those—and it's by far the greatest majority of contemporary theologians—who believe that there are natural pieces of evidence for the existence of God or gods. And so I have no problems with any of that. And, as you say, if a god existed who gave us free will or merely noted that we had free will, and wished to let our free will operate, then he or she or it might very well give us no evidence of his, her, or its existence for just that reason.

And this is connected with one of the many little tangents in the extraterrestrial-intelligence problem. In fact, there is a perfect parallel between the two cases. Let me spend a moment on it. Two sorts of arguments have been generated. One says that if extraterrestrial intelligence exists, then it would have capabilities vastly in excess of our own. Look at what we've done in just a few thousand years of civilization. Imagine some other beings who are millions or thousands of millions of years more advanced than we. Imagine what they could do. Why aren't they here? Why haven't they so rearranged the cosmos so that their existence is apparent just by looking up at the night sky? "Drink Coca-Cola" spelled out in stars. Something of that sort. A more religious message than that. But why isn't the universe so clearly artificial that there would be no doubt of the existence of extraterrestrial intelligence? This is in no way a different argument; it's just recast in modern language in slightly different terms. And one of the explanations—there are large numbers of them; on an issue with no data it is possible to have very involved debates—one of the explanations is the so-called zoo hypothesis, which says that there is an ethic of noninterference with emerging civilizations, because the extraterrestrials wish to see what humans will do. Let them develop on their own without outside interference, and therefore there is a stringently adhered-to requirement that none of the advanced civilizations make planetfall on Earth. And it seems to me that's very similar, not identical, to what you were saying about omnipotence and free will.

Questioner: Concerning the point about God leaving some amazing piece of evidence in the scriptures of His existence: I think that God's purpose is to leave evidence through all time for all men, even children, to understand that He exists, not to leave one piece of evidence for somebody to discover in a thousand years that will benefit one generation.

CS: No, all generations subsequently.

Questioner: Or all generations subsequently, but—
CS: A thousand years is as an instant in Thy sight.

Questioner: As one day. Right. I don't believe as a physicist that physics deals with the truth. I believe that it deals with successive approximations to the truth.

CS: So do I.

Questioner: I think if it ever dealt with the truth, that we'd be out of a job. So I am aware in the history of physics that you can't say that you've got the definitive equation for gravity or the definitive equation for quantum mechanics or anything like this. And that reminds me, actually, of a quote from Einstein that says God doesn't play with dice. And I find that difficult to reconcile with the views that you put out for Einstein's assumption that God was equivalent to the universe and the laws of quantum mechanics.

CS: Surely that is consistent. All he was saying is that he believed there were hidden variables behind which the statistical regularities of quantum mechanics could be derived in the same sense that ordinary Newtonian mechanics could. That's all he said.

Questioner: Yes, but he was not accepting present-day quantum mechanics as being the end of the story.

CS: Right. He was saying that the indeterminacy of quantum mechanics conflicted with his sense of a universe ruled by physical laws.

Questioner: And he put that down to God.

CS: Which he called God. That's right.

Questioner: Thank you.

CS: But which is very different from the traditional kind of God.

Questioner: Well, it may or may not be.

CS: Einstein was explicit that it was different. For example, in his first visit to the United States, he was sent an anguished telegram by the archbishop of Boston wanting to know what exactly were his religious views. And he spelled them out very explicitly and very courageously, and there was no question that it was not the traditional religious view of God. I mean, it doesn't matter, because Einstein is just one man. But since we all admire him, it's good to know what he actually said.

Questioner: Yes.

CS: And it was not the traditional view at all.

Questioner: Yes, well, yes. I accept that. Talking about proofs for the existence of God, I'd like to put it in perspective that there's no completely satisfactory proof that everyone in this room exists. I don't know if you know of one. I think it comes down in the end to belief of one sort or another that people in this room exist, and putting the proofs about God's existence in that context, we're demanding a lot more in proving God's existence than we are in proving our own existence.

CS: But the burden . . . the burden of proof is on those who claim that God exists. Or do you think not?

Questioner: I think you say that. I don't think that, in fact.

CS: You think the burden of proof is on those who say that God does not exist?

Questioner: An equal burden of proof, I would say. I don't see why it should be put to those who say that He exists.

CS: But would you say that, no matter what contention is made, that the burden of proving or disproving it falls equally on those who agree and those who disagree?

Questioner: I *would* say that.

CS: Have you thought of the political applications of this?

Questioner: Well, it's not a political issue, I don't think.

CS: No, but I thought it was a general proposition you were proposing.

Questioner: If you take a physical proposition, would you say you know that in every case the burden of proof rests to prove one type of case or the other type of case?

CS: The burden of proof always falls on those who make the contention.

Questioner: Well, all right. Yes. But only in the sense that it's disproving the other contention.

CS: No, no. It can be in an area where no one has any other contentions.

Questioner: Yes, well . . .

CS: It is—and it seems to me quite proper. Because otherwise opinions would be launched very casually if those who proposed them did not have the burden of demonstrating their truth. Here is a set of thirty-one proposals that I make, and good-bye. I mean, you would be left with a chaotic circumstance.

Questioner: Yes, all right. Yes, I see. I see your point. Yes.

CS: The audience is laughing. May I say I think these are . . . some of these are very good points, and this sense of dialogue I welcome and find delightful.

Questioner: I didn't agree with the way you presented some of the proofs for the existence of God. There was one other proof that I would like to give. I wouldn't call it a proof. I'd call it an argument, because I don't believe that you can prove in absolute logical terms the existence of God.

CS: So we are in agreement.

Questioner: There was an eminent scientist called Sir James Jeans, a Fellow of our Royal Society in the 1930s, who published a book called *The Mysterious Universe,* in which he went into great detail discussing the new discoveries of physics. And he presented a rather elegant argument concerning the existence of God, which was based on a very simple, almost unspoken law, the law being that if any two things interact, they must be in some way like. He then went on to say that it's quite possible for somebody who looks at the Sun at sunrise on a nice morning to have a beautiful, poetic thought about it. He looked at the chain of events, which went to producing that poetic thought. It started off in the Sun, with light being emitted, traveling across space, coming through the upper atmosphere, being refracted, and then eventually reaching the lens of the eye, being focused on the retina, and traveling as a nerve impulse to the brain, and then producing a thought.

Now, he said that there are two ways of looking at this. Either you can say that thought is a form of energy in some way, for its ability to interact with energy, or energy is a form of thought in some way.

CS: Those are two of a larger number of possible ways of looking at it.

Questioner: Two of a larger number. Yes. Now, scientists who restrict themselves to the purely rational view of man would say that, well, it's obvious, then, that thoughts are a form of energy.

CS: No, this is not a good argument. This is a 1930s premodern-neurology argument. "Thoughts are a form of energy."

Questioner: Well, it's equally valid to say that, you know, maybe the energy that's in the universe is in some way related to thought.

CS: They may be, perhaps, in some way related.

Questioner: If it is, for there to be one universe that everyone observes as being the same, there must be one being producing the thought.

CS: Why? Why? Why can't natural selection accommodate large numbers of unrelated organisms to the same laws of nature?

CHAPTER SEVEN

CS: I have a letter that I was sent that concluded by saying, "I have at times found your views somewhat naive and immature but hope for better things this week." I hope I have not disappointed. Let me read one remark of this deeply concerned person, who requested anonymity. He says, "On several occasions it has seemed to me that you try to quantify what is a qualitative experience. There is a spiritual and psychical world superimposed, as it were, on the physical. Worlds within worlds. Man is not just a physical being but a spiritual and a psychic entity, too."

Well, my only response is that this is a claim that, from my point of view, remains to be proved. I would have to ask, "What is the evidence that we are more than material beings?" I don't think anyone would doubt that matter is a part of our makeup. And the question is, what is the compelling evidence that it is not all?

Questioner: Sir, I have a feeling that we have a lot of growing to do. The scientist doesn't perhaps know yet how to bring a greater being into the picture, and suddenly there are psychic things that are spiritual. You're taking the wrong set of faculties to disprove the psychic element. You must use the similar faculty. So it will be hundreds of years before scientists can ever prove the psychic part of life.

CS: Would you grant the possibility that there is no psychic part of life?

Questioner: No.

CS: Not a possibility? Not a smidgen of doubt in your mind?

Questioner: I'm one of those who lives with one foot on each side of life. One foot on the psychic and a very practical other foot, as a businesswoman, on the world. I've proved it.

CS: What in general should we do in a dialogue like this? Here I am. I say that my mind is open. I am happy to see the evidence, and the response I sometimes get is, "I've had this experience. It's compelling to me. But I can't give it over to you." Now, doesn't that prevent any dialogue whatever? How are we to communicate?

Questioner: Well, you see, I think you're stopping with the mental faculties you have and saying, "This is me. This is wrong." Now, there are faculties that one could certainly not create, because they're already in the mind, spiritual faculties.

CS: Well, you see, I say they're not—that's not demonstrated—that there's no evidence that they exist. First you have to show that they exist before you can have a major program to encourage them.

Questioner: I don't know that you have to play the piano to know that you can.

CS: No. But I can require, at least, before I start practicing the piano that I see that a piano exists, that I see someone sit down at the piano, move his or her fingers, and produce music. That then convinces me that there is such a thing as a piano, there is such a thing as music, and it is not hopelessly beyond the ability of humans to produce music from a piano. But when I ask for something comparable in the psychic world, I am never shown it. I never have someone come up and produce an—I don't know—a twenty-foot-high psychic dragon. Or have some-one come and write down on the blackboard the demonstration of Fermat's last theorem. There simply is never anything that you can get your teeth into. You understand why I feel a little frustrated about this?

Questioner: I do. Yes. But then you possess faculties that can open that door to you.

CS: You're relying on *me* to find the psychic world? No.

Questioner: I'm hoping every individual can find it for themselves. It's a question of education within oneself.

CS: I believe that before we do the education, we have to first demonstrate that there is something to be educated on. I don't for a moment maintain that there isn't an enormous amount we have yet to learn. I believe that we have in fact dis-covered the tiniest fraction of the wonders of nature that are out there. But I just think until those who believe in the spiritual or psychic or whatever-you-want-to-call-it world can actually demonstrate in any way its existence, that it is not likely that sci-entists will be devoting a great deal of their time to adumbrat-ing this possibility.

Questioner: How dependable an evidence would you say is the electroencephalograph readings that have been taken in certain experiments on those who practice different types of meditation, perhaps from the Eastern teachings, and have been able to record more central brain-wave patterns during a time when the physical senses have been shut off and the mind has gone deep into the conscious, subconscious, unconsciousness if you like? That was done at Berkeley University [the Univer-sity of California, Berkeley] with a good friend of mine, where she was put into a simulated environment to create these circumstances.

CS: Well, I certainly agree that there is such a thing as the unconscious mind. There is all sorts of evidence for it in our everyday lives, and Freud provided a compelling argument that it exists. And I think it is essential that we understand it, and I believe that it plays a powerful, maybe even dominant, role in international relations, and that's therefore a very practical reason for understanding it.

I also believe that there are altered states of consciousness that can be brought about by some—it's related to what I said before—by sensory deprivation and by certain molecular assists. But I don't know of any evidence that it isn't a different mode of interaction of the molecules in our brain, a different sequence of flashing connections of neurons; that is, that there are other ways in which the brain works is guaranteed. That we don't fully understand those ways is also guaranteed. But that this is something other than matter—not a smidgen of evidence for that. Is that responsive?

Questioner: Yes it is.

CS: Thank you.

Questioner: Professor Sagan, this is a question on the God hypothesis. Don't you think that science, out of habitually having to find the answers for material things and having to be seen to attempt to find the answers, subject to public pressure and admiration, has ventured on this occasion into religious territory on which it should perhaps make a more cautious approach, in relation to your own admitted lack of scrupulous proof and unsubstantial faith? To my mind I thought science was a servant of mankind and not mankind a servant of science.

CS: I certainly agree with the last sentence, but I don't see how that is connected with the rest of what you said. My personal sense is that there are limitations, of course, to science, and I just indicated what a tiny fraction of the world I think we understand. But it is the only method that has been demonstrated to work. And if we bear in mind how liable we are to be deceived, to deceive ourselves—that was the point of some of the UFO discussions we had—then it is clear that what we need is a very hard-nosed and skeptical approach to contentions that are made in this area. And that hard-nosed and skeptical approach has been tested and honed, and it is called science.

"Science" is only a Latin word for "knowledge." And it's hard for me to believe that anyone is opposed to knowledge. I think that science works by a careful balance of two apparently contradictory impulses. One, a synthetic, holistic, hypothesis-spinning capability, which some people believe is localized in the right hemisphere of the cerebral cortex, and an analytic, skeptical, scrutinizing capability, which some people believe is localized in the left hemisphere of the cerebral cortex. And it is only the mix of these two, the generating of creative hypotheses and the scrupulous rejection of those that do not correspond to the facts, that permits science or any other human activity, I believe, to make progress.

As far as me bringing a scientific approach to the matters of religion, I think that is implicit in inviting a scientist to give the Gifford Lectures. I could hardly have left my science outside the door as I walked in. I would have appeared before you naked.

Questioner: Just at the end of your lecture, you referred to Bertrand Russell saying that you should not believe a proposition that you do not have good grounds for believing to be true.

Now, surely that in itself is a proposition. What grounds would you have for believing that proposition?

CS: Yes. That's a very good question that leads to an infinite regress. And notice that Russell said he would merely propose for our consideration this proposition. Russell was, in his mathematician incarnation, the author of precisely such logical paradoxes as the one you just suggested. So if you wish to have the statement justified in internal logic—that is, a self-consistent closed system—obviously it cannot, because it leads to an infinite regress. But as I was saying, it seems to me that the approach of skeptical scrutiny commends itself to our attention because it has worked so well in the past. So many findings—I tried to give some simple physical and astronomical ones in the earlier lectures—were made possible by science *not* accepting the conventional wisdom, *not* taking on blind faith what was taught in the religious and secular schools, that everybody knew—the teachings of Aristotle on physics and astronomy, for example—but instead by asking, "Is there really evidence for it?" It is the method of science. And at every step along the way, it has produced some agonizing reappraisals and some powerful emotions that don't like it. And I understand that very well. But it seems to me that if we are not dedicated to the truth in this sense of truth, then we are in very bad shape.

CHAPTER EIGHT

Questioner: How serious do you think the problem is with the creationists that are in the States?

CS: Well, different people will have a different answer. Some fundamentalist Christians believe that it is without any doubt that the world will end shortly, that the signs, especially the formation in 1948 of the state of Israel, are clear; that is, there are many fundamentalist Christians, at least in the United States—I don't know about elsewhere in the world—who deeply believe that this is true. And there will be a tribulation and a rapture, and there's an entire mythology about the events that will happen. We are even told by the Reverend Mr. Falwell that believing Christians, when the trumpet is sounded, will be taken bodily to heaven. And if they are driving a car or flying an airplane at that moment, then the car and airplane containing its nonbelieving passengers are in some difficulty. The conclusion of which would seem to be that there has to be a test of faith before issuing a license.

Questioner: You seem to think that in the event of a nuclear war, all human beings may become extinct. I put the question on the grounds of two things that you didn't bring up at all in your talk: One, nuclear power stations will be damaged in a nuclear war, and that will leak radiation that will be dangerous for thousands of years, and two, we don't know the effects of ultraviolet light that may come through to Earth after a nuclear war.

CS: Right. So the questioner says, is it clear that other forms of life would survive bearing in mind the enhanced ultraviolet flux from the destruction of the ozone layer and the radioactive fallout, especially if nuclear power plants are targeted. I chose grasses and cockroaches because of their high radiation resistance. And if you check it out, you find that they are several orders of magnitude more resistant than humans are. A typical dose of radiation to kill a human being is a few hundred rads. There are organisms that are not killed until a few million rads. Also, the sulfur-eating marine worms that I mentioned, they were not selected randomly either. They live entirely at the ocean bottom where no ultraviolet light can get and where they are quite well insulated against radioactivity in the environment. So for those reasons I still say that many forms of life would survive, and its clear from past mass extinctions like the Cretaceous-Tertiary event that many forms of life have survived in the past what were probably more serious events than a nuclear war, although it's quite true that the radioactivity was not a component of such events in the past.

Questioner: As a scientist, would you deny the possibility of water having been changed into wine in the Bible?

CS: Deny the possibility? Certainly not. I would not deny any such possibility. But I would, of course, not spend a moment on it unless there was some evidence for it.

CHAPTER NINE

CS: There was one question that was sent to me in a letter to my hotel, which was signed, "God Almighty." Probably just to attract my attention. It said that the writer's definition of a miracle would be if I would answer the letter. So to show that miracles can happen, I thought I would answer the question. The question was a straightforward and important one, often asked: "If the universe is expanding, what's it expanding into? Something that isn't the universe?"

Well, the way to think of this is to remember that we are trapped in three dimensions, which constrains our perspective (although there's not much we can do about being trapped in three dimensions). But let us imagine that we were two-dimensional beings. Absolutely flat. So we know about left/right and we know about forward/back, but we've never heard of up/down. It is an absolutely incoherent idea. Just nonsense syllables. And now imagine that we live on the surface of a sphere, a balloon, let's say. But of course we don't know about that curvature through that third dimension, because that third dimension is inaccessible to us, and we cannot even picture it. And now let's imagine that the sphere is expanding, the balloon is being blown up. And there is a set of spots on the balloon, each of which represents, let us say, a galaxy. And you can see that from the standpoint of every galaxy all the other galaxies are running away. Now, where is the center of the expansion?

On the surface of the balloon, the only part of it that the flat creatures can have access to, where is the center of the expansion? Well, it isn't on that surface. It's at the center of the balloon in that inaccessible third dimension. And, in the same way, into what is the balloon expanding? It is expanding in that perpendicular direction, that up/down direction, that inaccessible direction, and so you cannot, on the surface of the balloon, point to the place into which it is expanding, because that place is in that other dimension.

Now up everything one dimension and you have some sense of what people are talking about when they say that the universe is expanding. I hope that that was helpful, but considering the auspices of the writer, you should have known it anyway.

Questioner: A program from the Reagan administration was over the television last week. Mr. Paul Warnke stated that Star Wars [the Strategic Defense Initiative, or SDI] would fail.

CS: Well, maybe I should just say a few words about Star Wars. Star Wars is the idea that it's dreadful to be threatened with mass annihilation, especially at the hands of some people you've never met, and wouldn't it be much better to have an impermeable shield that protects you against nuclear weapons, to simply shoot down the Soviet warheads when they're on their way here? And as an idea it's an okay idea. The question is, can it be done? And let me not quote the legion of technical experts who believe that it is nonsense. Let me instead quote its most fervent advocates in the American administration, in the Department of Defense. *They* say that after some decades and the expenditure of something like one tr—Well, *they* don't actually say the expense, but it's an expenditure of something like one trillion dollars, that the United States might be able to shoot down between 50 and 80 percent of the Soviet warheads.

Let us imagine that the Soviet Union does nothing in the next few decades to improve its offensive capability; it leaves everything (a very unlikely possibility) at its present offensive force—that's ten thousand weapons. Ten thousand nuclear warheads. Let us give the benefit of the doubt to the exponents of Star Wars and imagine that instead of 50 to 80 percent they can shoot down 90 percent of the warheads. That leaves 10 percent that they cannot shoot down.

Ten percent of ten thousand warheads is (an arithmetical exercise accessible to everyone) one thousand warheads. One thousand warheads is enough to utterly demolish the United States. So what are we talking about?

The advocates say it can't protect the United States. And there are many other things that could be said about it, but I think that is a key point. Its advocates think it won't work. And it will cost a trillion dollars. Should we go ahead?

Questioner: Do you think that your people will go ahead?

CS: Why do something so foolish? A very good question. And here we are getting into murky issues of politics and psychology and so on, but I don't believe in ducking questions—I'll tell you what I think. I think that the alternative is abhorrent to the powers that be. The alternative is that you negotiate massive, verifiable, bilateral reductions in nuclear weapons, which would be an admission that the entire nuclear arms race has been foolish beyond belief, and that all of those leaders—American and Russian and British and French—for the last forty years, who bought this bill of goods put their nations at peril. It is such an uncomfortable admission that it takes great character strength to admit to it. So I think that rather than admit to it we are looking at a desperate attempt to have still more technology to get us out of the problem that the technology got us into in the first place. The ultimate technological fix. Or, as it is sometimes called, "the fallacy of the last move." Just one more ratchet up the arms race, please let us have it, and then everything will be fine forever. And if there's anything that's clear from the history of the nuclear arms race, it's that this isn't the case. Each side, generally the Americans, invents a new weapons system, and then the other side, generally the Soviets, invents it back. And then both nations are less secure than they were in the first place, but they've spent a charming amount of money and everybody's happy. Now, there's no question that if you wave a trillion dollars at the world aerospace community, you will have organizations, corporations, military officers, and so on interested in it, whether or not it will work.

And I'm sure that this is a part of it. But it's not the main part. The main part is a tragic reluctance to come to grips with the bankruptcy of the nuclear arms race. In the United States, it's eight consecutive presidents, something like that, of both political parties, that have bought it. Most of the people who run the country are advocates of the nuclear arms race, or have been in the past. It's very hard to say, "Sorry, we made a mistake," on an issue of this size. That's my guess.

Questioner: I think for the first time yesterday President Reagan offered to share the technology of SDI with the Russians.

CS: It's not the first time. He's been saying that all along.

Questioner: Yeah, but isn't it perhaps preferable that the joint efforts of the great powers be extended for perhaps defensive matters rather than the offensive weapons that have occupied them for so long?

CS: No, I don't agree. We're talking about a shield. Let's imagine another kind of shield, the contraceptive shield. Let's suppose that the contraceptive shield lets only 10 percent of the spermatozoa through. Is that better than nothing, or isn't it? I maintain that that's worse than nothing—among other things, for giving a false sense of security. But on the idea of sharing the technology, this is an administration that will not give an IBM personal computer to the Soviets. And we are asked to believe that the United States will hand over the eleventh-generation battle-management computer, which is decades off, and which will be so complicated that its program cannot be written by a human being or any collection of human beings. It can be written only by another computer. It cannot be debugged by any human being. It can be debugged only by another computer. And it can never be tested except in a nuclear war itself. And this we will hand over to the Russians? In either case, if we believed it would work or if we didn't believe it would work, I can't imagine the Russians saying, "Thank you very much. We will now have this as the principal mainstay of the security of the Soviet Union, this program that the Americans have very kindly just given over to us."

Nor can I imagine that the United States, after taking a sober look at this idea, would turn over the security of the country to this mad scheme. A system that has to work perfectly to protect the country and which can never be tested. Trust us. It'll be fine. Don't worry about it.

Questioner: Can religious beliefs adapt to the future?

CS: Well, it's certainly an important question. My feeling is, it depends on what religion is about. If religion is about saying how the natural world is, then to be successful it must adopt the methods, procedures, techniques of science and then become indistinguishable from science. By no means does it follow that that's all that religion is about. And I tried to indicate at the end of my last lecture some of the many areas in which religion could provide a useful role in contemporary society and where religions, by and large, are not. But that's very different from saying how the world is or came to be. And there the Judeo-Christian-Islamic religions have simply adopted the best science of the time. But it was a long time ago, the time of sixth-century B.C., during the Babylonian captivity of the Jews. That's where the science of the Old Testament comes from. And it seems to me important that the religions accommodate to what has been learned in the twenty-six centuries since. Some have, of course, to varying degrees; many have not.

Questioner: [*inaudible*]

CS: The god that Einstein was talking about is completely different, as I've tried to say several times in these lectures, from the standard Judeo-Christian-Islamic god. It is not a god who intervenes in everyday life, no microintervention, no prayer. It's not even clear that this god made the universe in the first place. So that's a very different use of the word "god" than what is, I gather, your attempt to justify the existing religion. That we have to use our sense organs and our intellectual abilities to comprehend these issues, I think, is apparent. Perhaps they are limited, but it's all we have. So do the best with what we have. Don't foist, I say, our predispositions on the universe. Look openly at the universe and see how it is. And how is it? It is that there's order in there. It's an amazing amount of order, not that we have introduced but that is there already. Now, you may choose to conclude from that fact that there is an ordering principle and that God exists, and then we come back to all the other arguments: Where did the ordering principle come from? Where did God come from? If you say that I must not ask the question of where God came from, then why must I ask the question of where the universe came from? And so on.

Questioner: Professor Sagan, I'd like advice, please. Is there anything you think an individual could do to change in some way the world situation, or should we just sit back and accept it?

CS: Nope, you don't have to sit back. I think if we let the governments do it, we will continue in the very desultory direction we have already been going for forty years or more. I think the first thing, in a democracy, where there is at least some pretense about the people controlling government policy, is that every democratic process ought to be used. You can make sure that those whom you vote for have rational views on these matters. You can work hard to make sure that there is a real difference of opinion in the alternative candidates. You can write letters to newspapers and so on. But more important than any of that, I believe, is that each of us must equip him- or herself with a "baloney-detection kit."

That is, the governments like to tell us that everything is fine, they have everything under control, and leave them alone. And many of us, especially on issues that involve technology, such as nuclear war, have the sense that it's too complicated. We can't figure it out. The governments have the experts. Surely they know what they're doing. They must be in favor of the support of our country, whichever our country happens to be. And anyway, this is such a painful issue that I want to put it out of my mind, which psychiatrists call denial. And it seems to me that that is a prescription for suicide, that we must, all of us, understand these issues, because our lives depend on them, and the lives of our children and our grandchildren. That's not an issue you want to take on faith. If ever there was a circumstance in which the democratic process ought to take hold, this is it. Something that determines our future and all that we hold dear. And therefore I would say that the first thing to do is to realize that governments, all governments, at least on occasion, lie. And some of them do it all the time—some of them do it only every second statement—but, by and large, governments distort the facts in order to remain in office.

And if we are ignorant of what the issues are and can't even ask the critical questions, then we're not going to make much of a difference. If we can understand the issues, if we can pose the right questions, if we can point out the contradictions, then we can make some progress. There are many other things that can be done, but it seems to me that those two, the baloney-detection kit and use of the democratic process where available, are at least the first two things to consider.

Questioner: [*inaudible*]

CS: Right. You say everyone in this room has felt aggression. Surely that's right. I'm sure it's right. There may be a few saints in the room . . . and I very much hope that there are. But at least almost everyone in the room must have felt it. But I also maintain that everyone in the room has felt compassion. Everyone in the room has felt love. Everyone in the room has felt kindness. And so we have two warring principles in the human heart, both of which must have evolved by natural selection, and it's not hard to understand the selective advantage of both of them. And so the issue has to do with which is in the preponderance. And here it is the use of our intellect that is central. Because we're talking about adjudicating between conflicting emotions. And you can't have an adjudication *between* emotions *by* an emotion. It must be done by our perceptive intellectual ability. And this is the place where Einstein said something very perceptive. In response—this is post–nuclear war, post-1945—in response to precisely the question you have just formulated, in which Einstein was saying that we must give the dominance to our compassionate side, he said, "What is the alternative?" That is, if we do not, if we cannot manage it, it is clear that we are gone. We're doomed. And therefore we *have* no alternative. Certainly untrammeled, continuing aggression in an age of nuclear weapons is a prescription for disaster. So either get rid of the nuclear weapons or change what passes for social relations among humans.

But even getting rid of nuclear weapons altogether will not solve this problem. There will be new technical advances. And already there are chemical and biological weapons that could perhaps rival some of the effects of nuclear war. So this is a very key aspect of what I was thinking when I said we are at a branch point in our history, in the sense of who we are. I maintain it's not a question of sudden change, that we have been compassionate for a million years, and it's a question of which part of the human psyche the governments—and the media, and the churches, and the schools—give precedence to. Which one do they teach?

Which one do they encourage? And all I'm saying is that it is within our capability to survive. I don't guarantee it. Prophecy is a lost art. And I don't know what the probabilities are that we will go one way or another. And no one says it's easy. But it is clear, as Einstein said, that if we do not make a change in our way of thinking, all is lost.

A c k n o w l e d g m e n t s

E diting these lectures afforded me, for precious moments at a time, the happy delusion that I was working with Carl once again. The words he spoke in these lectures would sound in my head and it felt wonderfully as if we had somehow been transported back to the two heavenly decades when we thought and wrote together.

We had the pleasure of writing several of our projects, the *Cosmos* television series among them, with the astronomer Steven Soter, our dear friend. Since Carl's death Steve and I wrote the first two planetarium shows for the magnificent Rose Center at the American Museum of Natural History in New York City. Once I had turned Carl's Gifford Lectures into a book, I invited Steve to join me in editing the final drafts. We felt sure that Carl would not have wanted us to use the 1985 slides from the lectures. Astronomers have seen farther and more clearly since then. Steve found the gorgeous images that replace them. He also wrote the scientific updates that appear in the footnotes. I am grateful to him for his many editorial contributions to this book.

Ann Godoff has been our editor ever since *Shadows of For-gotten Ancestors,* Carl's favorite among all the books he and we ever wrote. She also edited Carl's *Pale Blue Dot, The Demon-Haunted World,* and *Billions & Billions.* It was her recognition that the Gifford Lectures should become a book that made *The Varieties of Scientific Experience* possible. Her imagination and wit made the process of that transformation a pleasure. I thank her colleagues at the Penguin Press, art director Claire Vaccaro, and Ann's assistant Liza Darnton for all they did for the book and for me. I am grateful to Maureen Sugden for her meticulous and thoughtful copyediting.

Jonathan Cott has always been a North Star to me, guiding me to every possible kind of great cultural experience. I am fur-ther indebted to him for the valuable editorial comments and suggestions he gave me for this book.

I thank Sloan Harris of ICM, for his excellent representation and his consistent commitment to my work, and Katharine Clu-verius, in his office, for her kind assistance.

Kristin Albro and Pam Abbey in my office at Cosmos Studios have provided valuable administrative support, and Janet Rice helped in a host of ways, making it possible for me to focus on this work.

I wish to acknowledge the encouragement and loving kind-ness of Harry Druyan, Cari Sagan Greene, Les Druyan and Viky Rojas-Druyan, Nick and Clinnette Minnis Sagan, Sasha Sagan, Sam Sagan, Kathy Crane-Trentalancia, and Nancy Palmer.

Carl's Gifford Lectures were expertly transcribed from au-diotapes long ago by Shirley Arden, his executive assistant at the time. As I read the transcripts, which were done without the text-processing magic of today's computer technology, I felt a renewed sense of respect for her consistently meticulous work.

I would also like to thank the organizers of the Gifford Lectures and the University of Glasgow for their kind invitation to Carl and their hospitality to us during our time in Scotland.

In the ten years since Carl's death, these lectures sat in one of the thousand drawers of his vast archives. For some reason the Gifford Lectures were never logged into the archives' otherwise reliably comprehensive index. In the midst of a worldwide pandemic of extreme fundamentalist violence and during a time in the United States when phony piety in public life reached a new low and the critical separation of church and state and public classroom were dangerously eroded, I felt that Carl's perspective on these questions was needed more than ever. I searched in vain for the transcripts. Our friend, who wishes to remain anonymous, succeeded where I had failed. My gratitude to him for this, and much else, is profound.

· ANN DRUYAN
Ithaca, New York
March 21, 2006

Figure Captions

A 2004 image of Comet NEAT made by the Gunma Observatory of Japan. Every little red/green/blue dash is the spectral trace of a star.

FRONTISPIECE: HUBBLE ULTRA-DEEP FIELD

In 2004 the Hubble Space Telescope looked at a small piece of sky (a tenth the size of the full Moon) for eleven days to make this image of nearly ten thousand galaxies. Light from the most distant galaxies took almost thirteen billion years to travel the distance to Hubble's lens. Each galaxy contains many billions of stars, each star a potential sun to perhaps a dozen worlds.

Science lifts the curtain on a tiny piece of night and finds ten thousand galaxies hidden there. How many stories, how many ways of being in the universe are contained therein? All residing in what, to us, had been just a little patch of empty sky.

Figure 1. EAGLE NEBULA

A stellar nursery located about 6,500 light-years away from us. Through a window in a dark enveloping shell of interstellar dust, we see a cluster of brilliant newborn stars. Their intense blue light has sculpted filaments and walls of gas and dust, clearing and illuminating a cavity in a cloud about 20 light-years across.

Figure 2. CRAB NEBULA

This is the remnant of the same exploded star, or supernova, that Chinese and Native American Anasazi astronomers observed in the constellation Taurus in A.D. 1054. They recorded the sudden appearance of a brilliant new star that then slowly faded from view. The filaments are the unraveling debris of the star, enriched in heavy elements produced by the explosion.

Figure 3. SUN AND PLANETS

Here in their order and relative sizes are the Sun (at left), the four terrestrial planets (Mercury, Venus, Earth, Mars), the four gas giant planets (Jupiter, Saturn, Uranus, Neptune), and Pluto (far right).

Figure 4. WRIGHT SOLAR SYSTEM AND SIRIUS

The top shows to scale the Sun (left) and the orbit of Mercury (right). The middle shows the entire solar system with the orbit of Saturn (S) and several elliptical comet orbits (left) and the system of the bright star Sirius (right). The bottom shows from left to right the orbits of Saturn, Jupiter, Mars, Earth, Venus, Mercury, and the Sun.

Figure 5. SOLAR SYSTEM SCALES

Upper left: The orbits of the inner planets Mercury, Venus, Earth, and Mars, the asteroid belt, and the orbit of Jupiter.

Upper right: The scale increases tenfold to encompass the larger orbits of all the gas giant planets Jupiter, Saturn, Uranus, and Neptune, and the elliptical orbit of Pluto.

Lower right: A further scale change compresses the orbits of all the planets into the box at one end of the highly elliptical orbit of a comet.

Lower left: The scale increases again so that the cometary orbit is now in the tiny box at the center and we see the inner portion of the Oort Cloud of comets.

Figure 6. Oort Cloud

Schematic view shows the vast spherical cloud of perhaps a trillion comets, weakly bound by the gravity of the Sun (center). It was named after the Dutch astronomer Jan Oort, who correctly hypothesized its existence in 1950.

Figure 7. Wright: Other Systems

Wright imagined that our own solar system was but one of a countless number of similar systems in the Milky Way, each perhaps containing a star surrounded by its own retinue of planets and comets.

Figure 8. The Pleiades Star Cluster

The bright stars in this cluster illuminate the faint remnants of the interstellar cloud from which they formed. This star cluster, a naked eye object in the constellation Taurus, is about 13 light-years across.

Figure 9. Orion Nebula

A vast cloud of glowing interstellar gas and opaque dust, which is giving birth to dozens of new stars. The nebula is about 40 light-years across and 1,500 light-years away. If you look up at the constellation Orion on a winter night, this stellar nursery appears as the hazy central "star" in his sword.

Figure 10. Eskimo Nebula

Ten thousand years ago this halo of gas and dust was part of the central star. The aging star then expelled its outer layers into space in successive bursts, forming what astronomers call a planetary nebula. All ordinary stars like the Sun will eventually meet a similar fate.

Figure 11. VEIL NEBULA

These glowing filaments trace a portion of the expanding remnants of a supernova, a star that exploded about five thousand years ago in the constellation Cygnus.

Figure 12. SAGITTARIUS STAR CLOUD

A relatively crowded region of old stars in the direction of the center of the Milky Way Galaxy.

Figure 13. ANDROMEDA GALAXY, M31

This large spiral galaxy is only about 2 million light-years away, making it the closest one to our own Milky Way. The flattened rotating disk of stars and clouds of gas and dust is about 200,000 light-years across and contains several hundred billion solar systems.

Figure 14. HERCULES CLUSTER

Most of the objects in this image are entire galaxies, like our own Milky Way, each containing many billions of stars. Many of the galaxies of the Hercules Cluster are interacting, with some of them actually colliding and merging. This rich cluster is about 650 million light-years away.

Figure 15. SATURN WIDE SHOT

A stunning array of orbiting rings encircles the gas giant planet Saturn, which casts its shadow on them. The Cassini Division is the most prominent of many gaps in the ring system. It is named after the seventeenth-century Italian-French astronomer Giovanni Domenico Cassini who made many important discoveries about our solar system. His namesake spacecraft, the one that took this picture, has now done the same.

Figure 16. CLOSE-UP OF SATURN'S RINGS

In this back-lit image from the Cassini spacecraft, the Sun illuminates Saturn's rings from behind, revealing the fine structure of multiple thin rings.

Figure 17. SOLAR NEBULA

A chaotic cloud of interstellar gas and dust collapses under its own gravity (A). Most of the mass falls to the center to form and ignite the Sun, but the residual spin of the cloud prevents it from collapsing in one direction, resulting in a flat rotating disk (B). The particles in the disk coagulate to form larger objects, and the largest ones sweep out clear lanes from the debris disk (C). This process continues as the colliding particles become larger and fewer (D), eventually leaving the solar system in its present form (E).

Figure 18. PLANETESIMALS

In this stage of formation of a planetary system, colliding asteroid-size bodies orbit around the central star.

Figure 19. BETA PICTORIS

This 1997 false-color image shows a debris disk seen edge on in orbit around the star Beta Pictoris, which some twenty years earlier had provided the first evidence of planetary formation around a star outside our solar system. The telescope has blocked out the direct light from the star to reveal the fainter light reflected from the disk. The inner gap in the disk suggests that planets are forming there. Most young stars have such orbiting disks.

Figure 20. COMET MACHHOLZ

The extended atmosphere, or coma, of the comet blows away from the Sun to form faint tails of dust and ionized gas.

Figure 21. OLIVE OIL AND COMETS

English astronomer William Huggins compared the spectra of vaporized olive oil and ethylene (olefiant gas) with the spectra of two comets, which he observed in 1868. He correctly deduced that comets contain carbon-bearing substances.

Figure 22. COMET NEAT SPECTRUM

The light of Comet NEAT (shown on the jacket of this book) is spread out into its constituent rainbow of colors (bottom), revealing the presence of different molecules at particular wavelengths (middle).

Figure 23. END OF THE WORLD

An illustration by R. Jerome Hill, published in *Harper's Weekly*, May 14, 1910, depicting the romantic fatalism inspired by the coming of the "cyanide laden" Halley's Comet.

Figure 24. IAPETUS

The surface of this mysterious satellite of Saturn has two distinct zones, one icy and very bright, the other covered by a very dark red material of unknown composition. This bimodal distribution of brightness is unique in the solar system, as is the ridge around the satellite's equator.

Figure 25. SATURN SMALL MOONS

The satellites shown here range in size from about 20 to 200 kilometers. They lack sufficient gravity to enforce a spherical shape.

Figure 26. URANUS RINGS

This infrared image, taken at a wavelength of 2.2 microns, reveals several distinct rings encircling the planet. The isolated bright spot is the moon called Miranda.

Figure 27. PHOBOS

This curiously potatolike cratered inner moon of Mars has an average diameter of 22 kilometers and an orbit period of about eight hours.

Figure 28. DEIMOS

The outer moon of Mars has an average diameter of 13 kilometers and an orbit period of thirty hours.

Figure 29. MARS SURFACE BY *VIKING 1*

The view from the *Viking 1* Lander on the surface of Mars, in 1977, shows a rocky landscape and a ruddy sky. The lander in the foreground has its meteorology arm extended.

Figure 30. TITAN DISK

The largest moon of Saturn, with its intriguing features photographed by the Cassini orbiting space probe in 2005.

Figure 31. TITAN COAST

Showing icy highlands with dry rivers and what appears to be the shoreline of a vanished sea, as seen by the Huygens descent probe from an altitude of about 10 kilometers in 2005.

Figure 32. SAGITTARIUS STARS

The Spitzer Space Telescope turned its gaze toward the constellation Sagittarius. Its infrared camera was able to penetrate the obscuring curtains of gas and dust for a thrilling look at the crowded center of the Milky Way Galaxy.

Figure 33. SETI SPECTRUM

A graph of the natural radio background noise over a wide range of frequencies. At lower frequencies (left), charged particles in our galaxy emit increasing noise. At higher frequencies (right), the intrinsic quantum noise of any radio receiver increases. Between them is a relatively quiet "window," where interstellar hydrogen (H) and hydroxyl (OH) emit radio energy at discrete frequencies. This plot does not include radio emission from molecules in the Earth's atmosphere.

Figure 34. SIMULATED SETI SIGNAL

The search for extraterrestrial intelligence includes the monitoring of stars at many radio frequencies simultaneously over time. A successful detection might resemble this signal, which actually came from the *Pioneer 10* spacecraft in the outer solar system. The drift in frequency over time shows that the source is not rotating with the Earth, but is of extraterrestrial origin.

Figure 35. THE CRETACEOUS-TERTIARY RECORD IN THE ROCKS AT GUBBIO

The evidence for the event that caused the extinction of the dinosaurs sixty-five million years ago was discovered in this sequence of sedimentary strata from Gubbio, northern Italy. The pale limestone layers at the lower right were deposited in the Cretaceous period, when dinosaurs ruled the Earth. The darker limestone layers at the upper left are from the subsequent Tertiary period, when they had become extinct. In between, the diagonal layer of black clay contains the worldwide iridium-rich fallout of debris from the crater excavated by the collision of an asteroid or comet. This layer is found everywhere on Earth where rocks of this age are exposed. The edge of a coin at the top is for scale.

Figure 36. CRETACEOUS-TERTIARY IMPACT

Don Davis, one of the greatest painters of science-based art, takes us to the panicky last second of the age of the dinosaurs. An asteroid or comet some 10 kilometers in diameter plunged through the shallow ocean near what is now Yucatán in Mexico, igniting global wildfires and producing a dense cloud of smoke and dust that darkened and froze the surface of the Earth.

Index

Credits